UCLA Symposia on Molecular and Cellular Biology, New Series

Series Editor
C. Fred Fox

Volume 1
Differentiation and Function of Hematopoietic Cell Surfaces, Vincent T. Marchesi and Robert C. Gallo, *Editors*

Volume 2
Mechanisms of Chemical Carcinogenesis, Curtis C. Harris and Peter A. Cerutti, *Editors*

Volume 3
Cellular Recognition, William A. Frazier, Luis Glaser, and David I. Gottlieb, *Editors*

Volume 4
Rational Basis for Chemotherapy, Bruce A. Chabner, *Editor*

Volume 5
Tumor Viruses and Differentiation, Edward M. Scolnick and Arnold J. Levine, *Editors*

Volume 6
Evolution of Hormone-Receptor Systems, Ralph A. Bradshaw and Gordon N. Gill, *Editors*

Volume 7
Recent Advances in Bone Marrow Transplantation, Robert Peter Gale, *Editor*

Volume 8
Gene Expression, Dean H. Hamer and Martin J. Rosenberg, *Editors*

Volume 9
Normal and Neoplastic Hematopoiesis, David W. Golde and Paul A. Marks, *Editors*

Volume 10
Mechanisms of DNA Replication and Recombination, Nicholas R. Cozzarelli, *Editor*

Volume 11
Cellular Responses to DNA Damage, Errol C. Friedberg and Bryn A. Bridges, *Editors*

Volume 12
Plant Molecular Biology, Robert B. Goldberg, *Editor*

Volume 13
Molecular Biology of Host-Parasite Interactions, Nina Agabian and Harvey Eisen, *Editors*

Volume 14
Biosynthesis of the Photosynthetic Apparatus: Molecular Biology, Development and Regulation, J. Philip Thornber, L. Andrew Staehelin, and Richard B. Hallick, *Editors*

Volume 15
Protein Transport and Secretion, Dale L. Oxender, *Editor*

Volume 16
Acquired Immune Deficiency Syndrome, Michael S. Gottlieb and Jerome E. Groopman, *Editors*

Volume 17
Genes and Cancer, J. Michael Bishop, Janet D. Rowley, and Mel Greaves, *Editors*

Volume 18
Regulation of the Immune System, Harvey Cantor, Leonard Chess, and Eli Sercarz, *Editors*

Volume 19
Molecular Biology of Development, Eric H. Davidson and Richard A. Firtel, *Editors*

Volume 20
Genome Rearrangement, Ira Herskowitz and Melvin Simon, *Editors*

Volume 21
Herpesvirus, Fred Rapp, *Editor*

Volume 22
Cellular and Molecular Biology of Plant Stress, Joe L. Key and Tsune Kosuge, *Editors*

Volume 23
Membrane Receptors and Cellular Regulation, Michael P. Czech and C. Ronald Kahn, *Editors*

Volume 24
Neurobiology: Molecular Biological Approaches to Understanding Neuronal Function and Development, Paul O'Lague, *Editor*

Volume 25
Extracellular Matrix: Structure and Function, A. Hari Reddi, *Editor*

UCLA Symposia Published Previously

(Numbers refer to the publishers listed below.)

1972
Membrane Research (2)

1973
Membranes (1)
Virus Research (2)

1974
Molecular Mechanisms for the Repair of DNA (4)
Membranes (1)
Assembly Mechanisms (1)
The Immune System: Genes, Receptors, Signals (2)
Mechanisms of Virus Disease (3)

1975
Energy Transducing Mechanisms (1)
Cell Surface Receptors (1)
Developmental Biology (3)
DNA Synthesis and Its Regulation (3)

1976
Cellular Neurobiology (1)
Cell Shape and Surface Architecture (1)
Animal Virology (2)
Molecular Mechanisms in the Control of Gene Expression (2)

1977
Cell Surface Carbohydrates and Biological Recognition (1)
Molecular Approaches to Eucaryotic Genetic Systems (2)
Molecular Human Cytogenetics (2)
Molecular Aspects of Membrane Transport (1)
Immune System: Genetics and Regulation (2)

1978
DNA Repair Mechanisms (2)
Transmembrane Signaling (1)
Hematopoietic Cell Differentiation (2)
Normal and Abnormal Red Cell Membranes (1)
Persistent Viruses (2)
Cell Reproduction: Daniel Mazia Dedicatory Volume (2)

1979
Covalent and Non-Covalent Modulation of Protein Function (2)
Eucaryotic Gene Regulation (2)
Biological Recognition and Assembly (1)
Extrachromosomal DNA (2)
Tumor Cell Surfaces and Malignancy (1)
T and B Lymphocytes: Recognition and Function (2)

1980
Biology of Bone Marrow Transplantation (2)
Membrane Transport and Neuroreceptors (1)
Control of Cellular Division and Development (1)
Animal Virus Genetics (2)
Mechanistic Studies of DNA Replication and Genetic Recombination (2)

1981
Immunoglobulin Idiotypes (2)
Initiation of DNA Replication (2)
Genetic Variation among Influenza Viruses (2)
Developmental Biology Using Purified Genes (2)
Differentiation and Function of Hematopoietic Cell Surfaces (1)
Mechanisms of Chemical Carcinogenesis (1)
Cellular Recognition (1)

1982
B and T Cell Tumors (2)
Interferon (2)
Rational Basis for Chemotherapy (1)
Gene Regulation (2)
Tumor Viruses and Differentiation (1)
Evolution of Hormone-Receptor Systems (1)

Publishers

(1) Alan R. Liss, Inc.
41 East 11th Street
New York, NY 10003

(2) Academic Press, Inc.
111 Fifth Avenue
New York, NY 10003

(3) W.A. Benjamin, Inc.
2725 Sand Hill Road
Menlo Park, CA 94025

(4) Plenum Publishing Corp.
227 W. 17th Street
New York, NY 10011

NEUROBIOLOGY:
Molecular Biological Approaches to Understanding Neuronal Function and Development

NEUROBIOLOGY: Molecular Biological Approaches to Understanding Neuronal Function and Development

Proceedings of a UCLA Symposium
Held in Keystone, Colorado
April 1–7, 1984

Editor

PAUL O'LAGUE

Biology Department
University of California
Los Angeles, California

Alan R. Liss, Inc. • New York

Address all Inquiries to the Publisher
Alan R. Liss, Inc., 41 East 11th Street, New York, NY 10003

Printed in the United States of America.

Library of Congress Cataloging in Publication Data

Main entry under title:

Neurobiology—molecular biological approaches to understanding neuronal function and development.

Includes bibliographies and index.
1. Neurobiology—Congresses. 2. Developmental neurology—Congresses. 3. Molecular biology—Congresses.
I. O'Lague, Paul. II. University of California, Los Angeles.
[DNLM: 1. Cytoplasmic Filaments—congresses. 2. Ion Channels—congresses. 3. Molecular Biology—congresses. 4. Neurons—growth and development—congresses.
5. Neurons—physiology—congresses. W3 U17N v. 24/
WL 102.5 N4925 1984]
QP351.N4276 1985 612.8 85-4539
ISBN 0-8451-2623-7

Pages 1–91 of this volume are reprinted from the Journal of Cellular Biochemistry, Volumes 26 and 27. The Journal is the only appropriate literature citation for the articles printed on these pages. The page numbers in the table of contents, contributors list, and index of this volume correspond to the page numbers at the foot of these pages.

The table of contents does not necessarily follow the pattern of the plenary sessions. Instead, it reflects the thrust of the meeting as it evolved from the combination of plenary sessions, poster sessions, and workshops, culminating in the final collection of invited papers, submitted papers, and workshop summaries. The order in which articles appear in this volume does not follow the order of citation in the table of contents. Many of the articles in this volume were published in the Journal of Cellular Biochemistry, and they are reprinted here. These articles appear in the order in which they were accepted for publication and then published in the Journal. They are followed by papers which were submitted solely for publication in the proceedings.

Contents

Contributors

Michael A. Alejos, Department of Cell Biology, University of Texas Health Science Center, Dallas, TX 75235 **[55]**

Ruth Hogue Angeletti, Laboratory of Neuropathology, Department of Pathology and Laboratory Medicine, University of Pennsylvania School of Medicine, Philadelphia PA 19104 **[31]**

Hirohiko Aoyama, Institut d'Embryologie du CNRS et du College de France, 94130 Nogent sur Marne, France **[39]**

Rene Arentzen, Central Research & Development Department, E.I. du Pont de Nemours and Company, Wilmington, DE 19898 **[65]**

Frank Baldino, Jr., Central Research & Development Department, E.I. du Pont de Nemours and Company, Wilmington, DE 19898 **[65]**

R.L. Barchi, Institute of Neurological Sciences and the Department of Neurology, University of Pennsylvania School of Medicine, Philadelphia, PA 19104 **[9]**

Martin W. Berchtold, Department of Cell Biology, Baylor College of Medicine, Houston, TX 77030; present address: Universitat Zurich-Irchel, Institut fur Pharmakologie und Biochemie, Arbeitsgruppe Biochemie, Winterhurerstrasse 190, CH-8057 Zurich, Switzerland **[169]**

Mordechai Bodner, Departments of Hormone Research and Organic Chemistry, The Weizmann Institute of Science, Rehovot, Israel **[21]**

Daniel F. Bowen-Pope, Department of Pathology, SM-30, University of Washington, Seattle, WA 98195 **[145]**

Martin Carden, Laboratory of Neuropathology, Department of Pathology and Laboratory Medicine, University of Pennsylvania School of Medicine, Philadelphia, PA 19104 **[31]**

Michael Comb, Department of Chemistry, University of Oregon, Eugene, OR 97403; present address: Department of Molecular Biology, Massachusetts General Hospital, Boston, MA 02114 **[179]**

Martha Constantine-Paton, Department of Biology, Princeton University, Princeton, NJ 08544; present address: Biology Department, Yale University, New Haven, CT 06511 **[151]**

Erica J. Cox, Department of Cell Biology, University of Texas Health Science Center, Dallas, TX 75235 **[55]**

Leonard G. Davis, Central Research & Development Department, E.I. du Pont de Nemours and Company, Wilmington, DE 19898 **[65]**

James Douglass, Department of Chemistry, University of Oregon, Eugene, Oregon 97403 **[179]**

Jean Loup Duband, Institut d'Embryologie du CNRS et du College de France, 94130 Nogent sur Marne, France **[39]**

Mati Fridkin, Department of Organic Chemistry, The Weizmann Institute of Science, Rehovot, Israel **[21]**

The number in brackets is the opening page number of the contributor's article.

Jonathan A. Fuller, Department of Neurobiology and Behavior, SUNY at Stony Brook, Stony Brook, NY 11794 [127]

R.E. Furman, Institute of Neurological Sciences and the Department of Neurology, University of Pennsylvania School of Medicine, Philadelphia, PA 19104 [9]

Joel M. Gottesfeld, Department of Molecular Biology, Research Institute of Scripps Clinic, La Jolla, CA 92037 [93]

Illana Gozes, Department of Hormone Research, The Weizmann Institute of Science, Rehovot, Israel [21]

Karen F. Greif, Department of Biology, Bryn Mawr College, Bryn Mawr, Pennsylvania, PA 19010 [1]

W. Sue T. Griffin, Department of Cell Biology, University of Texas Health Science Center, Dallas, TX 75235 [55]

Ashley T. Haase, Infectious Disease Section, Veterans Administration Medical Center, San Francisco, CA 94121; present address: Department of Microbiology, University of Minnesota, Minneapolis, MN 55455 [121]

Edward Herbert, Department of Chemistry, University of Oregon, Eugene, OR 97403 [179]

Gerald A. Higgins, Central Research & Development Department, E.I. du Pont de Nemours and Company, Wilmington, DE 19898 [65]

James E. Krause, Department of Neurobiology and Behavior, SUNY at Stony Brook, Stony Brook, NY 11794; present address: Department of Anatomy and Neurobiology, Washington University School of Medicine, St. Louis, MO 63110 [127]

Norbert E. Kremer, Department of Neurobiology and Behavior, SUNY at Stony Brook, Stony Brook, NY 11794 [127]

J.F. Kuo, Department of Pharmacology, Emory University School of Medicine, Atlanta, GA 30322 [101]

Raymond J. Lasek, Department of Developmental Genetics and Anatomy, Case Western Reserve University School of Medicine, Cleveland, OH 44106 [135]

Elias Lazarides, Division of Biology, California Institute of Technology, Pasadena, CA 91125 [73]

Virginia M.-Y. Lee, Laboratory of Neuropathology, Department of Pathology and Laboratory Medicine, University of Pennsylvania School of Medicine, Philadelphia, PA 19104 [31]

Yuan Lin, Central Research & Development Department, E.I. du Pont de Nemours and Company, Wilmington, DE 19898 [65]

Dane Liston, Department of Chemistry, University of Oregon, Eugene, OR 97403 [179]

Robert W. Manning, Central Research & Development Department, E.I. du Pont de Nemours and Company, Wilmington, DE 19898 [65]

Mitchel Martin, Department of Chemistry, University of Oregon, Eugene, OR 97403 [179]

Jeffrey F. McKelvy, Department of Neurobiology and Behavior, SUNY at Stony Brook, Stony Brook, NY 11794 [127]

Anthony R. Means, Department of Cell Biology, Baylor College of Medicine, Houston, TX 77030 [169]

Robert J. Milner, Department of Molecular Biology, Research Institute of Scripps Clinic, La Jolla, CA 92037 [93]

Marcelle R. Morrison, Department of Neurology, University of Texas Health Science Center, Dallas, TX 75235 [55]

W. James Nelson, Division of Biology, California Institute of Technology, Pasadena, CA 91125 [73]

Monica M. Oblinger, Department of Developmental Genetics and Anatomy, Case Western Reserve University School of Medicine, Cleveland, OH 44106; present address: Department of Biological Chemistry and Structure, The Chicago Medical School, North Chicago, IL 60064 [135]

Paul O'Lague, Biology Department, University of California, Los Angeles, CA 90024 [xvii]

Thomas J. Poole, Institut d'Embryologie du CNRS et du College de France, 94130 Nogent sur Marne, France **[39]**

John A. Putkey, Department of Cell Biology, Baylor College of Medicine, Houston, TX 77030 **[169]**

De-Fang Qi, Department of Pharmacology, Emory University School of Medicine, Atlanta, GA 30322; present address: Shanghai Institute of Biochemistry, Chinese Academy of Sciences, Shanghai, People's Republic of China **[101]**

Thomas A. Reh, Department of Biology, Princeton University, Princeton, NJ 08544; present address: Department of Medical Physiology, University of Calgary, Alberta T2N 4N1 Canada **[151]**

Wanda Reynolds, Department of Molecular Biology, Research Institute of Scripps Clinic, La Jolla, CA 92037 **[93]**

Michael A. Rogawski, Laboratory of Neurophysiology, National Institute of Neurological and Communicative Disorders and Stroke, National Institutes of Health, Bethesda, MD 20205; present address: Department of Neurology, The John Hopkins Hospital, Baltimore, MD 21205 **[113]**

Russell Ross, Department of Pathology, SM-30, University of Washington, Seattle, WA 98195 **[145]**

William W. Schlaepfer, Laboratory of Neuropathology, Department of Pathology and Laboratory Medicine, University of Pennsylvania School of Medicine, Philadelphia, PA 19104 **[31]**

Audrey Seasholtz, Department of Chemistry, University of Oregon, Eugene, OR 97403 **[179]**

Ronald A. Seifert, Department of Pathology, SM-30, University of Washington, Seattle, WA 98195 **[145]**

Yael Shani, Department of Hormone Research, The Weizmann Institute of Science, Rehovot, Israel **[21]**

Rosalia C.M. Simmen, Department of Cell Biology, Baylor College of Medicine, Houston, TX 77030 **[169]**

Kim D. Stewart, Department of Neurobiology and Behavior, SUNY at Stony Brook, Stony Brook, NY 11794 **[127]**

J. Gregor Sutcliffe, Department of Molecular Biology, Research Institute of Scripps Clinic, La Jolla, CA 92037 **[93]**

J.C. Tanaka, Institute of Neurological Sciences and the Department of Neurology, University of Pennsylvania School of Medicine, Philadelphia, PA 19104 **[9]**

Jean Paul Thiery, Institut d'Embryologie du CNRS et du College de France, 94130 Nogent sur Marne, France **[39]**

Holly I. Trenchard, Department of Biology, Bryn Mawr College, Bryn Mawr, Pennsylvania PA 19010 **[1]**

John Q. Trojanowski, Laboratory of Neuropathology, Department of Pathology and Laboratory Medicine, University of Pennsylvania School of Medicine, Philadelphia, PA 19104 **[31]**

Gordon C. Tucker, Institut d'Embryologie du CNRS et du College de France, 94130 Nogent sur Marne, France **[39]**

R. Scott Turner, Department of Pharmacology, Emory University School of Medicine, Atlanta, GA 30322 **[101]**

Michel Vincent, Institut d'Embryologie du CNRS et du College de France, 94130 Nogent sur Marne, France; present address: Département de Génétique Moléculaire et Humaine, Centre Hospitalier de l'Université Laval, Saint Foy, G1V 4G2 Canada **[39]**

Jeffrey D. White, Department of Neurobiology and Behavior, SUNY at Stony Brook, Stony Brook, NY 11794 **[127]**

Betty Wolfson, Central Research & Development Department, E.I. du Pont de Nemours and Company, Wilmington, DE 19898 **[65]**

Preface

One of the fastest growing areas of modern biology is the molecular biology of gene structure and function. This growth has its origins in techniques that now allow isolation, duplication, and amplification of genetic information and fairly rapid sequencing of long DNA stretches. The amount of information accrued with the help of these techniques over the last five years is nothing short of staggering. On the other hand, one of the last areas likely to be unravelled is that associated with uncovering molecular rules governing the development and organization of neurons into hierarchical synaptic networks responsible for mediating animal behavior and its modification with experience. Topics encompassing both these areas were the focus of a UCLA Symposium held in Keystone, Colorado in April of 1984. This meeting brought together an international group of neurobiologists and molecular biologists, all of whom willingly participated in informal exchange of information. From the intensity of exchange, we felt the meeting was a great success.

The papers in this volume are representative of the many topics discussed at the meeting, which included cytoskeletal architecture, receptors and growth factors, gene expression and multigene families, membrane channels, intracellular messengers, neuropeptides and transmitter plasticity, and organizational principles in the development of neuronal networks. The present volume is divided into four sections. The first contains papers on cell adhesion, intercellular and virus-cell interactions in the nervous system. Recent observations are presented on how migratory and adhesive properties govern the formation of peripheral ganglia, on transsynaptic regulation of extracellular matrix components, and on how synaptic interactions affect retinotectal maps. This section concludes with an important paper on viruses and the nervous system in which recent autoradiographic techniques of general interest to both molecular biologists and neurobiologists are presented. The second section has articles dealing with recent observations on cytoskeletal architecture. This highly focused section contains papers describing the use of monoclonal antibodies directed against specific cytoskeletal proteins, in situ hybridization techniques for the mRNA of cytoskeletal elements, evidence for membrane-cytoskeletal protein interactions, and regulation of neuronal cytoskeletal networks. The third section consists of papers on neuropeptides and growth factors and describes recent biochemical observations on neuropeptide regulation and growth factor functions. In addition, the papers give a glimpse at how the most up-to-date molecular biological techniques such as, for example, in situ hybridization, cDNA cloning, use of synthetic oligonucleotide probes, and DNA sequencing is being brought to bear on problems of brain chemistry. The fourth section concludes the volume with papers on intracellular messengers and ion channels. Problems dealt with here include molecular characteristics and functional reconstitution of excitable channels, electrophysiology of single channels, and phospholipid/Ca^{++}-dependent protein phosphorylation. The final paper in this section summarizes new molecular biological studies on calmodulin gene cloning, genes of great interest to neurobiologists because of their role in neuronal functioning. This volume should be useful for anyone interested in learning about some new directions in the study of the nervous system.

The program of this symposium was created with the advice and active participation of an Advisory Committee consisting of Richard Horn, Jim Patrick, Lewis Reichardt, Richard Scheller, and Greg Sutcliff. After deliberations by phone, we met at the Jerry Lewis Neuromuscular Research Center for a day of lively discussion in which the final details of program organization were thrashed out. After having made difficult decisions as to who should be invited, we then were faced with finding resources to help them come. Major gifts from Johnson and Johnson, McNeil Pharmaceutical Company, Sandoz, Ltd., E.I. du Pont de Nemours & Company, and Hoffman-La Roche, Inc. provided supplementary funding for travel and subsistence support for those invited meeting participants. All of us who attended are particularly grateful to Betty Handy for her efforts at the meeting which assured smooth operations, and to Sandy Malone and Hank Harwood whose work prior to the meeting made my part in it a pleasure rather than a chore.

Finally, I would like to thank Lynn Ianni, who coordinated editorial review by the Journal of Cellular Biochemistry for many of these articles, and who contributed the eleventh hour energy which helped close out this volume.

Paul O'Lague

Journal of Cellular Biochemistry 26:127–133 (1984)
Neurobiology: Molecular Biological Approaches to
Understanding Neuronal Function and Development 1–7

Evidence for Transsynaptic Regulation of Neuronal Cell Surface Heparan Sulfate Proteoglycan in Developing Rat Superior Cervical Ganglion

Karen F. Greif and Holly I. Trenchard

Department of Biology, Bryn Mawr College, Bryn Mawr, Pennsylvania 19010

The effect of neonatal deafferentation on the expression of a neuronal cell surface heparan sulfate proteoglycan (HeS-PG) was investigated in the developing rat superior cervical ganglion. Two monoclonal antibodies, one directed against the core protein of HeS-PG, and one to a determinant associated with a heparan sulfate side-chain, were used to monitor postnatal increases of HeS-PG by radioimmunoassay. Following neonatal deafferentation by section of the cervical sympathetic trunk, total protein per ganglion was slightly reduced at survival times of 7, 14, and 30 days. Expression of the core protein determinant on HeS-PG was not altered in deafferented ganglia. In contrast, levels of side-chain determinant were significantly reduced at 14 and 30 days. These results suggest that processing of HeS-PG side-chains by principal ganglionic neurons is partially regulated by transsynaptic influences during development. Transsynaptic regulation of neuronal development may be a more general process than was believed previously, with effects not limited to molecules associated with synaptic development.

Key words: **radioimmunoassay, superior cervical ganglion, heparan sulfate, transsynaptic regulation**

The rat superior cervical ganglion (SCG) has been used as a model system for studies of normal development and of the role of innervation in the regulation of development. Biochemical maturation of ganglionic neurons has been carried out by assay of tyrosine hydroxylase (TH), the rate-limiting enzyme in catecholamine biosynthesis [1,2]. Estimates of the progress of synaptogenesis by cholinergic preganglionic neurons have been carried out by assay of choline acetyltransferase (CAT) [1,3,4]. Morphological assessment of principal ganglionic neurons and of developing synapses has been carried out by electron microscopic analysis [1,5,6]. These studies have shown that the SCG completes its development postnatally, with increases in TH levels [1], and increases in ganglionic cell size. Recognizable synapses are rare in the SCG in neonatal animals, with synaptogenesis occurring most rapidly during the first two postnatal weeks [1,3,6].

Received April 19, 1984; accepted June 18, 1984.

Because of the ease of manipulation of its presynaptic input, the SCG has been used as a model in which to study transsynaptic effects of presynaptic input. Section of the cervical sympathetic trunk (CST) in neonatal rats affects both the pattern of neuronal maturation and that of synaptogenesis [1–3,6–8]. Isolation of the SCG from developing presynaptic input at birth results in reduction in the postnatal increase in TH levels and reduces cell division and growth [1–3]. The development of postsynaptic specializations associated with synaptogenesis also does not occur [6]. Since denervation can be mimicked by treatment with ganglionic blockers such as chlorisondamine and pempidine [4], it has been hypothesized that presynaptic input exerts its developmental effects via direct synaptic activity. However, additional trophic influences may contribute to postsynaptic changes associated with denervation [9].

If reinnervation is permitted, up to 50% of normal levels of synapses can be recovered by 60 days [8]. However, TH levels do not recover [10]. In an apparent paradoxical result, assay by high-performance liquid chromatography (HPLC) of catecholamine levels present in both deafferented and reinnervated ganglia do not differ from those measured in control ganglia [10]. Thus, despite persisting reductions in TH levels, catecholamine pools remain at control levels.

Recently, monoclonal antibodies have been used to monitor the development of a neuronal cell surface heparan sulfate proteoglycan (HeS-PG) in the SCG [11]. Antibodies directed against two different determinants on HeS-PG, one associated with the core protein and one with a heparan sulfate side chain, were used to assess neuronal maturation. Quantitation of antigen levels by radioimmunoassay (RIA) indicated that these antigens increased significantly during postnatal development, with the most dramatic increases occurring during the second postnatal week. No significant redistribution of antigen associated with development was observed. These antigens were of interest because of the association of HeS-PG with a factor that induces neurite outgrowth in vitro [12,13]. The antigen can be used as a marker for neuronal surface to monitor neurite outgrowth and increases in cell diameter. Alterations in expression may be associated with the regulation of neurite outgrowth.

Because HeS-PG may represent a molecule associated with more general characteristics of development than neurotransmitter synthetic enzymes, we were interested in examining whether the expression of determinants on both core protein and side chains is regulated by presynaptic input. If the levels of these molecules are also reduced by neonatal deafferentation, it may be possible to conclude that transsynaptic regulation of development is a factor not limited to the control of molecules directly associated with synaptic transmission. A preliminary report of a portion of this work has been published [14].

METHODS

Reagents

Monoclonal antibodies directed against HeS-PG were obtained from Drs. L.F. Reichardt and W.D. Matthew. Monoclonal antibody PG 3 recognizes an antigenic determinant associated with a heparan sulfate side chain, and PG 22 recognizes the core protein or an N-linked carbohydrate associated with it. These antibodies were used either diluted directly from culture supernatants or from ammonium sulfate precipitations of ascites fluid. Previous studies have indicated that no differences in the properties of antibodies derived from these sources can be observed [11]. ^{125}I-goat-anti-mouse-IgG (Fab′ fragment) or ^{125}I-sheep-anti-mouse-IgG (whole molecule) were obtained from Amersham.

Experimental Animals and Surgical Procedures

Sprague-Dawley rats were used throughout for these experiments. For neonatal section of the CST, published procedures were followed [6,8]. Litters of pups aged 1–4 days were removed from their mothers. The sex of each pup was determined and equal numbers of males and females were selected for experimental and control groups. Since unilateral deafferentation of the SCG can cause compensatory changes in the contralateral ganglion (Smolen, personal communication), all surgery was carrried out bilaterally. Control rats were littermates of experimental animals in all cases. Experimental animals were anesthesized with methoxyfluorane in air. The skin over the throat was opened and the carotid artery was visualized. The SCG was located beneath the carotid artery and the CST severed at the base of the ganglion. A 2–3-mm section of the CST was removed to prevent reinnervation. Section of the CST was carried out bilaterally. The skin was sealed with collodion dissolved in acetone. Pups were kept warm until wide-awake and were returned as a group, along with control animals, to their mother. Mortality was less than 20% for 7-day survival periods, with somewhat poorer survival rates with longer survival periods.

Surgery on adult female rats was carried out using chloral hydrate (35 mg/kg, 7% solution) as anesthetic. Bilateral section of the CST was carried out in a similar fashion. The fascia surrounding the salivary glands and the overlying skin were sutured separately using 4-0 silk. Rats were given intramuscular injections of 100,000 units of Bicillin to reduce postsurgical respiratory infections. Mortality was less than 5% using prophylactic doses of antibiotic.

Quantitation of Antigen

Assay of antigen levels was carried out according to previously developed methods [11]. Ganglia from experimental animals were pooled to provide sufficient tissue for assay. Control ganglia were treated similarly. Desheathed ganglia were homogenized in 5 mM Tris-HCl, pH 8.1, in the presence of a protease inhibitor, phenylmethylsulfonylfluoride. Protein levels were determined by Amido-Schwarz assay. The total protein per ganglion was determined by correcting this value for total volume of homogenate and the number of ganglia pooled. Twelve dilutions of whole tissue homogenate were prepared (from 1:1 to 1:10,000) and aliquots were incubated with equal volumes of antibody at limiting dilution. After 18–24 hr, bound antibody was precipitated by centrifugation at 100,000g in an Airfuge. The amount of antibody remaining in the supernatant was determined by a standard solid-phase RIA [15], using a crude preparation of rat brain synaptosomes as the antigen. 50% inhibition points (I-50) were determined graphically, and standardized to 4 mg/ml initial concentration of protein. These values were further corrected for variations in protein content per ganglion, to permit comparison between experimental and control groups at each age examined, according to I-50 (corrected) = I-50 × [total protein (control ganglia)/ total protein (experimental ganglia)]. Control results were standardized to a value of 1.0; results from experimental groups were calculated as antigen levels relative to control levels (relative specific activity, RSA). All homogenates were assayed at least twice.

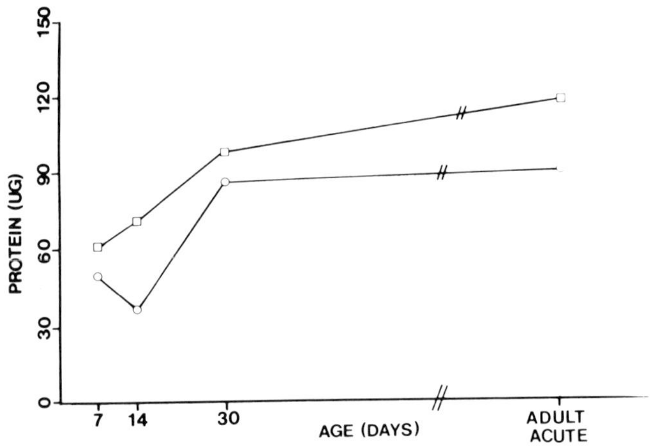

Fig. 1. Protein content of control and deafferented ganglia. Values determined as described in the text from homogenates of pooled SCGs. Open squares = control SCGs; open circles = deafferented SCGs.

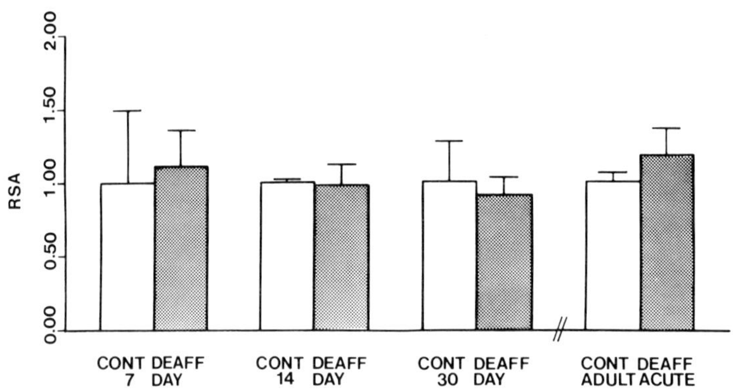

Fig. 2. Radioimmunoassay of levels of core-protein determinant recognized by PG 22. Control values for each age (in days) have been standardized to an average value of 1.0. Values for deafferented groups are presented as fractions of control levels (RSA = I-50 deaff/ I-50 control for each age group). Bars = one standard error of the mean. See text for discussion.

RESULTS

Protein assays of pooled experimental and control ganglia were carried out using tissue from 7-, 14-, and 30-day-old animals. In all cases, protein content of deafferented ganglia was lower than that of control groups (Fig. 1). The difference in ganglionic protein content was greater in younger animals than in 30-day groups, where a 12% reduction in total protein per ganglion was observed. After bilateral section of the CST and 7-day survival in adult rats, a 26% reduction in ganglionic protein content was observed, indicating loss of protein associated with the degeneration of presynaptic terminals (Fig. 1).

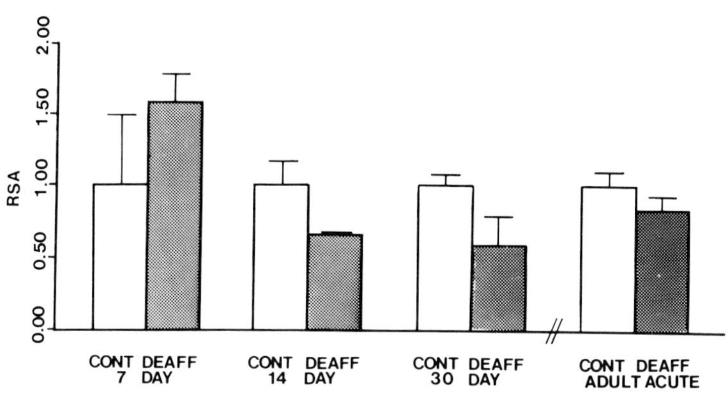

Fig. 3. Radioimmunoassay of levels of heparan sulfate side-chain determinant recognized by PG 3. Data presented as in Figure 2. See text for details.

RIA of levels of antigenic determinants recognized by PG 22 and PG 3 at 7, 14, and 30 days are shown in Figures 2 and 3. No significant differences in antigen levels recognized by PG 22 were observed at any time point following neonatal section of the CST, after correction for reduction in protein content per ganglion in experimental groups. In addition, no changes in levels of antigen recognized by PG 22 were observed following acute section of the CST in adult animals. In contrast, expression of the determinant recognized by PG 3 was significantly reduced at 14 and 30 days after birth (pooled, t-test, $P < 0.01$). Levels were increased over control levels at 7 days, but this difference was not significant. A small reduction in antigen level was also observed after acute section of the CST in adult rats, but this reduction was not statistically significant. Despite the reduction in levels of the side-chain determinant recognized by PG 3, absolute levels of the determinant increased during postnatal development of deafferented ganglia.

Immunocytochemical staining of experimental and control ganglia using either PG 3 or PG 22 followed by horseradish peroxidase-conjugated goat-anti-mouse antibody yielded no marked differences in staining pattern or staining intensity in 14 or 30 day SCG (data not shown).

DISCUSSION

The results of RIA of antigens associated with HeS-PG provide evidence that presynaptic innervation plays a role in the regulation of expression of molecules other than enzymes associated with neurotransmitter synthesis. This finding is important because it demonstrates that molecules that undergo significant changes associated with more general features of neuronal maturation [cf 11] can be influenced by innervation of the cell during critical postnatal periods, possibly mediated by activity [cf 4]. Innervation may play a central role in controlling the maturation of neurons, as has been demonstrated to be the case not only for development and maintenance of the neuromuscular junction but also for muscle tissue [cf 16].

Reductions in total protein per ganglion following neonatal deafferentation are consistent with previous studies [2,10]. SCGs deprived of preganglionic innervation from birth contain less protein at all ages examined, with the greatest reduction from control levels at 14 days. Much of the protein normally contributed by presynaptic axons and terminals is at least partially replaced by other sources, such as glia and connective tissue.

Studies of antigens associated with the HeS-PG indicate that the expression of a core protein antigen recognized by PG 22 does not appear to be regulated by presynaptic input. After correction for reduction in ganglion size, the same relative amount of antigen is synthesized in ganglia which have been deprived for presynaptic input from birth. In contrast, the pattern of synthesis of a heparan sulfate side-chain determinant, recognized by PG 3, is altered. At 7 days, more side-chain antigen is present in denervated SCGs. The variability in results was high for assay of 7-day tissue; further assays are required to determine whether this finding is significant. By 14 days, however, and persisting at 30 days, a 30–40% decrease in the level of side-chain antigen is observed. This result must be considered in view of earlier studies of normal development of these two antigenic determinants [11]. In these earlier studies, the relative amount of antigen recognized by PG 3 was less at birth than that of PG 22. In addition, immunocytochemical staining using PG 3 could not be detected until 7–10 days after birth, while staining using PG 22 was present at birth. These results suggested that postnatal processing of the side-chain determinant bound by PG 3 was required.

The results of the current study are compatible with this hypothesis, and further suggest that synthesis of the side-chain determinant may be partially regulated by presynaptic input. The absence of alterations in antigen levels following section of the CST in adult rats is consistent with reports of little, if any, immediate effect of deafferentation on ganglionic neurons in adult animals [cf 17].

Postnatal alterations in the metabolism of heparan sulfate proteoglycans occur in the rat brain [18], particularly during the first 2 weeks, and are proposed to be involved in postnatal neuronal maturation processes involving synaptogenesis. An embryonic-to-adult conversion of carbohydrate side-chains occurs in a neuronal cell adhesion molecule (N-CAM), in which loss of sialic acid residues enhances cell adhesion [19]. This alteration may be important in the modelling of the developing nervous system by regulating cell-cell interactions. In the ciliary ganglion, the ability of cultured neurons to respond to a neurite outgrowth factor is affected by their developmental stage; cells lose their responsiveness to a conditioned-medium factor after contact with target neurons is made [20].

The HeS-PG has been reported to be part of a complex that is a potent inducer of neurite outgrowth in vitro [12]. A mutant of the PC-12 cell line lacking cell surface HeS-PG also fails to secrete the neurite outgrowth factor. Our current observation that neonatal deafferentation appears to reduce levels of a side-chain determinant on HeS-PG may reflect a developmentally significant conversion of the molecule. However, the antibodies employed in the present study do not block the in vitro effect. Recently, antibodies have been generated to the neurite outgrowth factor that are capable of blocking its in vitro function [21]. It remains to be determined whether these antibodies exert a similar effect in vivo, and whether presynaptic input contributes to the regulation of expression of these antigenic determinants. Further investigation of HeS-PG is required to determine whether alterations in the molecule play a role in control of neurite outgrowth in vivo.

ACKNOWLEDGMENTS

We thank Dr. Leonard L. Ross for generously providing resources for research, and Dr. Arnold J. Smolen for valuable discussions and technical advice. Portions of this research were conducted at the Department of Anatomy, Medical College of Pennsylvania. Support was provided by a grant to K.F.G. from the Dysautonomia Foundation, Inc.

REFERENCES

1. Black IB, Hendry IA, Iversen LL: Brain Res 34:229, 1971.
2. Black IB: Annu Rev Neurosci 1:183, 1978.
3. Black IB, Hendry IA, Iversen LL: J Neurochem 19:1367, 1972.
4. Black IB, Geen SC: Brain Res 63:291, 1973.
5. Eranko L: Brain Res 46:159, 1972.
6. Smolen AJ, Raisman G: Brain Res 181:315, 1980.
7. Black IB, Coughlin MD, Couchard P: Soc Neurosci Symp 4:184, 1979.
8. Smolen AJ: Dev Brain Res 1:49, 1981.
9. Hendry IA, Hill CE: Brain Res 200:201, 1980.
10. Smolen AJ, Lindley T, Wright LL: Soc Neurosci Abstr 9:935, 1983.
11. Greif KF, Reichardt LF: J Neurosci 2:843, 1982.
12. Matthew WD, Lander AD, Greenspan RJ, Reichardt LF: Soc Neurosci Abstr 8:299, 1982.
13. Lander AD, Fujii DK, Gospodarowicz D, Reichardt LF: J Cell Biol 94:574, 1982.
14. Greif KF, Trenchard HI: J Cell Biochem (Suppl) 8B:116, 1984.
15. Klinman NR: J Exp Med 136:241, 1972.
16. Jacobsen M: "Developmental Neurobiology." New York: Plenum, 1978.
17. Gabella G: "Structure of the Autonomic Nervous System." London: Chapman and Hall, 1976.
18. Margolis R, Margolis RK, Chang LB, Preti C: Biochemistry 14:85, 1975.
19. Edelman GM: Science 219:450, 1983.
20. Collins F, Lee MR: J Neurosci 2:424, 1982.
21. Matthew WD, Patterson PH: Soc Neurosci Abstr 9:206, 1983.

Journal of Cellular Biochemistry 26:135–146 (1984)
Neurobiology: Molecular Biological Approaches to
Understanding Neuronal Function and Development 9–20

Molecular Characteristics and Functional Reconstitution of Muscle Voltage-Sensitive Sodium Channels

R.L. Barchi, J.C. Tanaka, and R.E. Furman

Institute of Neurological Sciences and the Department of Neurology, University of Pennsylvania School of Medicine, Philadelphia, Pennsylvania 19104

Action potentials in mammalian nerve and muscle are generated by a stereotyped sequence of alterations in membrane conductance to monovalent cations [1,2]. The fundamental change that leads to the initiation of an action potential is a regenerative, but self-limited, increase in membrane conductance to sodium ions [3]. This time- and voltage-dependent conductance pathway is controlled by a unique transmembrane protein, the voltage-sensitive sodium channel [4].

Detailed kinetic analysis of currents flowing through the sodium channel began more than 30 years ago with the pioneering work of Hodgkin and Huxley [3]. Biochemical characterization of this channel protein was undertaken more recently but has moved forward rapidly with concurrent advances in general membrane biochemistry. It is now possible to isolate the sodium channel protein from nerve and muscle membranes [5–7], to examine its molecular properties [8–10], and to reconstitute it into defined lipid vesicles [11–13].

In this report, the molecular properties of the voltage-sensitive sodium channel purified from mammalian skeletal muscle will be reviewed and the evidence for its functional integrity considered. Although this discussion will concentrate on recent progress that has been made with reconstitution of the sodium channel from skeletal muscle, reviews of biochemical approaches to sodium channels in general [14] and in rat brain in particular [15] have appeared elsewhere.

NEUROTOXINS AND SODIUM CHANNELS

Since the sodium channel has no natural ligands and performs no function in the absence of an intact membrane, identifying the solubilized channel protein can be a limiting factor in biochemical studies. Work with the channel, therefore, has relied heavily on its interaction with specific neurotoxins to locate this protein after its

Received April 1, 1984; accepted April 9, 1984.

removal from the membrane. One class of commonly used neurotoxins includes the small polar molecules tetrodotoxin (TTX) and saxitoxin (STX); these can be radiolabeled to high specific activity and have proven especially useful as reversible labels of the channel in biochemical studies [16]. A second group of nonpolar alkaloid neurotoxins, which includes batrachotoxin and veratridine, has been helpful in work with the reconstituted protein since these toxins can activate the channel without requiring a change in membrane potential [17]. Finally, a number of polypeptide neurotoxins from scorpions and from sea anemonae bind to the channel at one or more additional sites where they alter channel activation and inactivation and interact allosterically with the alkaloid neurotoxins [17]. All three groups of neurotoxins have been used extensively in biochemical studies of sodium channel proteins.

In its native membrane environment the voltage-sensitive sodium channel is characterized functionally on the basis of the ionic currents that it controls. The significance of results with the isolated channel protein depends heavily on the degree to which this protein retains these normal functional properties after purification. Although direct measurement of current gating by the purified channel is technically difficult, many of functional properties can be studied indirectly through the use of these neurotoxins. Initially, preservation of the specific binding sites for the various classes of toxins can be documented [11,18]. More importantly, these toxins can be used to open or close the purified channel after reconstitution into defined lipid vesicles so that channel function can be studied in the absence of a vesicle membrane potential [11–13].

ISOLATION OF MUSCLE SODIUM CHANNELS

The voltage-sensitive sodium channel, as identified by specific binding of tetrodotoxin and saxitoxin, has been isolated from rat and rabbit skeletal muscle by our group [6,19] and from rat brain and eel electroplax by others [5,7]. The muscle channels are most easily removed from the membrane with non-ionic detergents such as Nonidet P-40 (NP-40) or Triton X-100; ionic detergents such as CHAPS, deoxycholate, and cholate are effective in solubilizing the channel protein but are not as efficient in preserving its native conformation as measured by high affinity STX binding.

As was first demonstrated with the eel electroplax [20], phospholipid is required for sodium channel stability during purification [6]. When the channel is initially removed from the membrane, this phospholipid requirement can be satisfied by endogenous membrane lipid, but subsequent manipulation of the channel requires the addition of exogenous lipid to all buffers. For the muscle channel, phosphatidylcholine, mixed muscle phospholipids, or crude phospholipid extracted from soybean (asolectin) have proved to be effective protective agents. Ratios of detergent to phospholipid of 7:1 are sufficient to preserve channel function. In addition the high-affinity saxitoxin binding used to identify the channel protein in solution is temperature-labile after solubilization; all stages of channel purification must be carried out at 0–4°C.

Purification of the muscle sodium channel is accomplished by sequential chromatography of the solubilized membrane proteins on an ion exchange resin and a lectin affinity matrix, followed by centrifugation on a sucrose density gradient [6,9]. For ion exchange chromatography a weak ion exchange resin is prepared that has a guanidinium group at the terminus of a 20-atom hydrophilic spacer arm [6,9]; this resin retains the sodium channel while allowing >95% of the solubilized membrane protein to pass through unretarded. The peak of STX binding from this column is applied directly to a second column containing immobilized wheat germ agglutinin (WGA) [6,21]. Channel bound to the WGA is then eluted with N-acetylglucosamine, concentrated, and applied to a 5–15% sucrose density gradient. After centrifugation, fractionation of this gradient yields a single peak of specific STX binding that coincides with the major peak of protein on the gradient.

All stages of muscle membrane isolation and sodium channel purification are carried out in the presence of a panel of protease inhibitors that includes EGTA (0.5 mM), phenylmethylsulfonylfluoride (PMSF) (0.1 μM), iodoacetamide (1 mM), o-phenanthroline (1 mM), leupeptin (1.0 μg/ml), and pepstatin (0.1 μg/ml). This protocol has been applied successfully to rat and rabbit skeletal muscle channels [6,9,19]; preliminary studies suggest that it is applicable to human muscle as well (R. Roberts and R. Barchi, unpublished data).

For both the rabbit and the rat channel, specific activities in the purified material can approach 3,000 pmol STX bound per mg protein. If a molecular weight of 320,000 is assumed for the channel [22], a theoretical maximum specific activity of 3,130 pmol per mg can be expected. Thus, in the best preparations, channel purity probably exceeds 90%.

PHYSICAL PROPERTIES OF THE SOLUBILIZED SODIUM CHANNEL

The sodium channels from rat, rabbit, and human skeletal muscle appear very similar in their physical properties after solubilization (Table I). Chromatography on an analytical column of Sepharose 6B indicates an apparent Stokes' radius of approximately 8.6 nm for each in their mixed micellar form with detergent and phospholipid. This value is comparable to that reported for the channels from rat brain [23] and from eel electroplax [5,24].

TABLE I. Molecular Characteristics of Purified Sodium Channels

Source	Stokes radius (Å)	Apparent S value (H_2O)	Calculated $S_{20,w}$	Channel MW (calculated)	Subunits on SDS-PAGE
Eel electroplax	90[a] 85[b]	8	—		260,000[a] 270,000[b]
Rat muscle	86	9.1	9.8	314,000	230,000[c] (260,000)[d] 43,000 38,000
Rabbit muscle	86	8.6	8.8	295,000	260,000 38,000
Human muscle	85	8.3	9.1	287,000	
Rat brain	80	11.3	12	316,000	260,000 38,000 37,000

[a]See reference 5.
[b]See reference 24.
[c]Apparent molecular weight of large subunit after isolation of sarcolemma.
[d]True molecular weight of large subunit as determined by immunoblots with monoclonal antibodies [26].

Sedimentation measurements for muscle sodium channels in water yield apparent S values between 8.4 and 9.1 when compared to standard proteins of known $S_{20,w}$ [19,22]. These values will be in error, however, owing to the presence of lipid and detergent in the mixed micelle with the channel protein. An estimate of the true sedimentation coefficient of the solubilized protein can be made by comparing values obtained in water with those obtained relative to the same standard proteins in sucrose gradients prepared in D_2O. This approach is subject to some error since estimates must be made of the actual partial specific volume of the protein; nevertheless, comparison of the results obtained for each of the muscle sources is revealing. In each case, a value of 0.82–0.84 g/ml is found for the partial specific volume of the mixed micelle containing the channel and a calculated $S_{20,w}$ between 8.8 and 9.8 is obtained. Based on these values, the estimated molecular weight for the channels themselves is between 278,000 and 320,000 daltons [19,22]. Again, these values are comparable to those reported for the rat brain sodium channel [23].

Immunological data also support a strong similarity among sodium channel proteins. Monoclonal and polyclonal antibodies have been prepared against the purified rat muscle sodium channel [25,26]. Using immunocytochemical techniques, we find that polyclonal antiserum against the rat channel cross-reacts with the surface membrane of human, mouse, and guinea pig skeletal muscle, indicating shared immunogenic determinants between these species [25]. This antiserum also reacts with sodium channels at the rat peripheral nerve node of Ranvier, underscoring the similarity between nerve and muscle sodium channels. A number of the monoclonal antibodies prepared against the rat muscle channel show cross-reactivity with the rabbit muscle channel and with preparations from rat brain, further supporting the concept of close structural similarity among these channel proteins [26].

SUBUNIT COMPOSITION

Both the rabbit and the rat muscle sodium channels contain a large glycoprotein subunit with anomalous migratory characteristics on SDS-PAGE as well as one or more smaller subunits [9,19]. In highly purified preparations of rabbit sodium channel, a diffuse band is seen that migrates with an apparent molecular weight of ~260,000; a single second subunit appears at 38,000 MW on these gels (Fig. 1). These two bands are progressively enriched during the purification process, and codistribute with the peak of STX binding on the final sucrose gradient. In our purest preparations of rabbit sodium channel, these two proteins account for more than 90% of the total protein detectable on the gels [19].

The purified rat sodium channel exhibits more complex behavior on SDS-PAGE. In channel preparations obtained from isolated sarcolemma, a large glycoprotein subunit is found that runs as a very diffuse band between ~160,000 and 220,000 MW on 7–15% gradient SDS-PAGE [9]. In addition, two smaller bands are seen. One is comparable in size to the rabbit small subuinit, with a MW of 38,000, while the second is about 43,000 MW. The apparent MW of the large glycoprotein in both rat and rabbit channels varies considerably depending on the acrylamide concentration used for electrophoretic analysis. Ferguson plots of relative mobility vs. gel concentation suggest a free solution mobility for this component of the rat channel that is nearly twice that seen with standard proteins [9].

Several monoclonal antibodies raised against the purified channel protein have been used to trace the large glycoprotein of the rat channel during the purification of

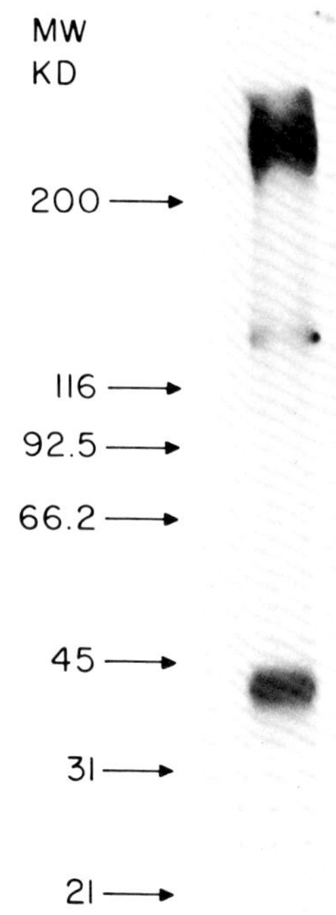

Fig. 1. Subunit composition of the purified sodium channel from rabbit skeletal muscle. The purified channel contains a large glycoprotein with an apparent MW of about 260,000 and a smaller component of 38,000 MW. Proteins were labeled with Bolton-Hunter reagent prior to electrophoresis, separated on a 7–20% gradient SDS-polyacrylamide gel, and visualized by autoradiography [9,19].

sarcolemma and the subsequent solubilization and purification of the sodium channel itself [26]. On blot transfers from 12% acrylamide-0.075% bis-acrylamide SDS-PAGE gels these antibodies recognize a diffuse, high molecular weight component between 160,000 and 200,000 MW among proteins from sarcolemmal membranes that have been prepared in the presence of a full panel of protease inhibitors; this component is detected throughout the various stages of the channel purification and corresponds to the large subunit of the purified channel. If muscle is homogenized in the presence of the same protease inhibitors, but the membranes are immediately solubilized by the inclusion of 1% NP-40, these same antibodies identify only a broad band centered at 260,000 MW [26]. This higher molecular weight band corresponds in location to the large glycoprotein in the purified rabbit channel. These antibody binding data suggest that the large subunit in the rat channel is also ~260,000 daltons in its native form but that modification by an endogenous protease or glycosidase occurs during preparation of sarcolemmal membranes in spite of the presence of protease inhibitors, producing a component that runs below 200,000 MW.

Large glycoprotein subunits have also been identified in each of the other sodium channels for which purification has been reported (Table I). In the rat brain [10] and eel preparation [8,24], this component has a MW of ~260,000, comparable to that obtained here for the rabbit channel and suggested for the rat channel by antibody binding [19,26]. The mammalian channels all contain at least one smaller component of 38,000 MW; no small subunits are seen in the channel preparations from eel electroplax.

The significance of the small subunits seen in the mammalian channel preparations remains to be defined. In spite of the presence of protease inhibitors in the muscle and brain preparations, it is possible that the 38,000 MW component is the result of proteolytic cleavage of the larger protein by endogenous enzymes. However, of twelve monoclonal antibodies that react with the large subunit, we have not found any that also cross-react with this smaller component as might occur if the latter was derived from the former [26]. Alternatively, the small protein may represent a component of the channel found only in higher species, or may be an associated protein derived from the cellular cytoskeletal system that assists in anchoring the sodium channel in mammalian excitable membranes. These points remain to be resolved.

RECONSTITUTION OF THE PURIFIED CHANNEL PROTEIN

The rabbit and the rat skeletal muscle sodium channels have been functionally reconstituted following purification [11,12,19]. In both cases, the channel protein isolated in the presence of NP-40 was reconstituted into egg phosphatidyl choline (PC) vesicles using BioBeads SM-2 to remove the detergent. This procedure results in the production of a bimodal population of vesicles; about half of the lipid is associated with vesicles 200–300 Å in diameter, while the rest is found in vesicles larger than 1,000 Å [11,13]. Freeze-fracture electron micrographs of vesicles containing the reconstituted rat muscle channel reveal intramembranous particles of ~95 Å diameter that are absent from control vesicles formed without the sodium channel [11].

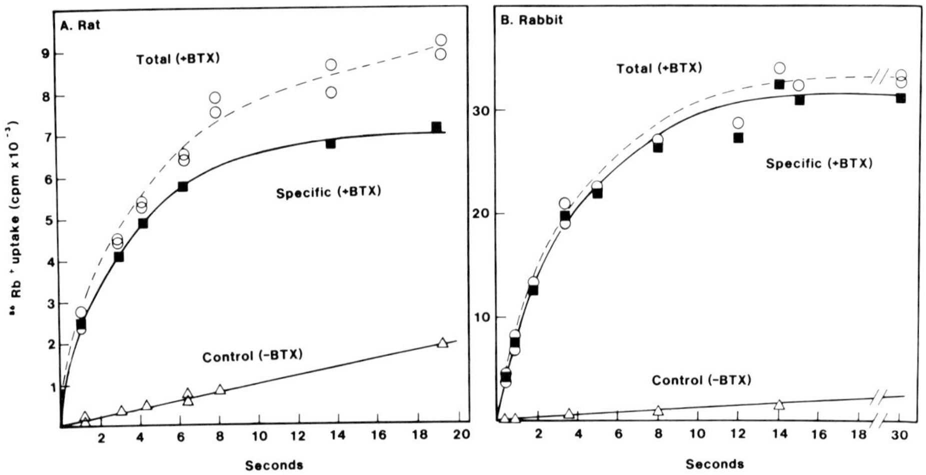

Fig. 2. Purified muscle sodium channels reconstituted into egg phosphatidyl choline (PC) vesicles gate cation fluxes in response to pharmacological activation. Vesicles containing purified rat channel protein (A) or rabbit channel protein (B) were equilibrated for 45 min with 1.5 μM batrachotoxin (○) or with buffer alone (△). Influx was initiated by rapidly mixing the activated vesicles with buffer containing $^{86}Rb^+$ and was terminated by injection of the mixture into a slurry of Dowex resin [12]. Rapid specific influx of $^{86}Rb^+$ (■; total influx with BTX minus control influx in the absence of BTX) was seen with both preparations. This influx was blocked ~50% by externally applied tetrodotoxin, and 100% when tetrodotoxin was present on both the inside and outside of the vesicle.

Measurements with [^{3}H]-saxitoxin indicate that about 50% of the high-affinity toxin binding sites are exposed on the external surface of the reconstituted vesicles. The remainder of these binding sites are evident only after disruption of vesicle integrity with detergents. Since the saxitoxin binding site is known to be on the extracellular side of the channel [16], these results imply that channel insertion during reconstitution is a random process. Similar results have been reported for the reconstituted sodium channel from rat brain [13].

Channel-containing vesicles that have been exposed to batrachotoxin, an alkaloid neurotoxin capable of opening the channel in the absence of a membrane potential [17], exhibit a very rapid influx of labelled cations that is not present in the absence of this toxin or in vesicles that do not contain sodium channels (Fig. 2) [11]. This influx is inhibited 50–70% by externally applied saxitoxin or tetrodotoxin. When the vesicles are formed with saxitoxin trapped inside, 100% of the remaining batrachotoxin-activated influx can be blocked by the subsequent application of external saxitoxin or tetrodotoxin. It appears that sodium channels inserted into the vesicle membranes either in an inward or outward facing direction can be activated by batrachotoxin, and that all the activated channels can be blocked by saxitoxin or tetrodotoxin if these agents are present on both sides of the membrane.

Batrachotoxin and veratridine activate the reconstituted channel in a dose-dependent manner (Table II). Our data suggest that both bind to a single class of sites with an apparent K_d of 1.5 μM and 35 μM, respectively [11]. Saxitoxin blocks the stimulated cation influx by binding to a separate site with a K_d of ~ 1 nM [12]. These values are all comparable to those measured with sodium channels in situ. Lidocaine and benzocaine, local anesthetics that inhibit sodium channel function in intact nerve and muscle, are also effective in blocking activation of the purified channel by batrachotoxin [27]. Flux studies with the reconstituted rat brain sodium channel have yielded comparable results with some of these toxins (Table II) [13].

TABLE II. Properties of Reconstituted Sodium Channels

Source	Rat muscle (sarcolemma)	Rabbit muscle (t-tubules)	Rat brain (synaptosomes)
Lipid	Egg PC	Egg PC	Egg PC
Channel orientation	50–70% outward	50% outward	50% outward
Channel activators	Batrachotoxin Veratridine Aconitine	Batrachotoxin Veratridine	Veratridine
Inhibitors of stimulated influx	Saxitoxin Tetrodotoxin Lidocaine Benzocaine	Saxitoxin Tetrodotoxin	Saxitoxin Tetrodotoxin Tetracaine
Cation selectivity	$Na^+ > K^+ > Rb^+ > Cs^+$ (Batrachotoxin and veratridine activation)	$Na^+ > K^+ > Rb^+ > Cs^+$ (Batrachotoxin and veratridine activation)	$Na^+ > Rb^+ > Cs^+$ (Veratridine activation)

QUENCHED FLOW MEASUREMENTS OF CATION SELECTIVITY

Due to the small size of the vesicles containing the reconstituted sodium channel, the cation fluxes stimulated by channel activators equilibrate very rapidly. In order to resolve the kinetics of cation influx, we adapted a quenched flow system to measure isotope uptake into activated vesicles [12]. This device is capable of mixing 20 to 50 μl of activated vesicles with an equivalent volume of buffer containing a labeled cation within 4 msec. After a variable holding period, cation uptake is quenched by injection of the mixture into a slurry of Dowex resin in the Tris form. The dead-time for the device operating in this mode has been determined to be 90 msec; after that interval, increments as small as 5 msec can be resolved.

Using this device, we have examined the rates of equilibration of four monovalent cations into vesicles containing the purified sodium channel from either rat or rabbit skeletal muscle [12,19]. After activation by batrachotoxin, uptake of Cs^+ proceeds rather slowly; at 22°C, the average half-time for vesicle equilibration is 10 sec. At the same temperature Rb^+ moves inward more rapdily, with an average half-time of 2.5 sec. Equilibration of K^+ is tenfold more rapid, with a half-time of 350 ms. Na^+ uptake is so fast that only the very last stages of vesicle filling are resolved; the process is virtually complete by 90 msec, and an upper limit of 50 msec can be placed on the half-time for this cation (Table III).

Based on the half-times for vesicle equilibration with each of these cations, a cation selectivity sequence and relative selectivity ratios can be calculated (Table III). These ratios confirm that the purified, reconstituted channel from both rat and rabbit muscle retain their selectivity for sodium over other cations [12,19]. These relative selectivity ratios are similar to those measured for the toxin-activated channel in its native membrane environment [28,29].

Measurements of cation uptake have also been made in vesicles activated with veratridine [12,19]. With this toxin, half-times for vesicle equilibration are several orders of magnitude slower than comparable measurements made with batrachotoxin-activated vesicles (Table III). It is necessary to raise the assay temperature to 37°C in order to produce influx rates sufficiently rapid to measure with adequate signal-to-noise ratio above the temperature-insensitive, nonspecific background influx. Under these conditions, the veratridine-activated channel shows much less apparent discrimination between cations; the half-times for Na^+ and K^+ are indistinguishable, while the slowest cation, Cs^+, requires less than tenfold longer periods for equilibration.

TABLE III. Cation Selectivity in the Reconstituted Sodium Channel

A. Rabbit T-tubular sodium channel

	Batrachotoxin							Veratridine						
Sequence	Na^+	>	K^+	>	Rb^+	>	Cs^+	Na^+	≃	K^+	>	Rb^+	>	Cs^+
T½	< 50 msec		390 msec		2.6 sec		6.5 sec	1.1 min		1.2 min		2.7 min		3.6 min
Ratio	1	:	0.13	:	0.019	:	.008	1	:	0.92	:	0.41	:	0.31

B. Rat sarcolemmal sodium channel

	Batrachotoxin							Veratridine						
Sequence	Na^+	>	K^+	>	Rb^+	>	Cs^+	Na^+	>	K^+	>	Rb^+	>	Cs^+
T½	< 50 msec		350 msec		2.5 sec		10 sec	1.0 min		1.2 min		2.0 min		2.6 min
Ratio	1	:	0.14	:	0.02	:	0.005	1	:	0.83	:	0.50	:	0.38

In order to shed more light on the difference in cation uptake rates through batrachotoxin- and veratridine-activated channels, measurements of the temperature dependence of cation influx have been made [12]. Rb^+ and Cs^+ influx following batrachotoxin activation show little temperature dependence; the a Q_{10} for cation influx is about 1.3 and the activation energy, calculated from Arrhenius plots, is ~ 7 kcal/mol for both cations. Measurements for the same cations after veratridine activation indicate a marked dependence on temperature; Q_{10} is measured to be about 3.0, with calculated activation energies of 31 kcal/mol for both cations.

Clearly the rate-limiting process for cation influx through batrachotoxin- and veratridine-activated channels is not the same. Electrophysiological [30] and recent patch clamp [31] data suggest that batrachotoxin-activated channels are open most of the time at 0 mV. Influx measurements in batrachotoxin-activated vesicles probably reflect the relative ease of cation movement through an open channel. Veratridine-activated channels, on the other hand, may flicker infrequently between the closed and opened state, spending much of their time closed [32]. The measured rates and temperature dependence of cation influx with this activator may reflect the probability of channel opening in a given vesicle rather than the actual cation selectivity of the open channel.

SINGLE-CHANNEL MEASUREMENTS

We have shown that purified, reconstituted muscle sodium channels retain the capacity to gate cation fluxes in response to pharmacological activation and that these fluxes can be specifically inhibited by appropriate neurotoxins. The purified channel still displays excellent cation selectivity. The next major feature that must be sought in the reconstituted channel is voltage activation of ionic currents. This is a more difficult issue to deal with since the time resolution of the quenched flow system is not sufficient to resolve events on the scale of the single-channel openings that occur in response to voltage [33]. We have approached this question using patch clamp of large liposomes.

Early work with single-channel recordings of purified sodium channels reconstituted into liposomes was hampered by artifacts produced by the detergents used in the channel purification and liposome preparation. Nonionic detergents usually used for sodium channel purification, such as NP-40, Triton X-100, and Lubrol PX, generated channel-like artifacts in patch-clamped lipid vesicles without protein, presumably because of traces of detergent that remain after BioBead treatment. These artifacts have been seen with various batches and brands of these detergents and do not seem to be an isolated problem with a single supplier (R. Furman, unpublished data). Bile salts, however, do not introduce these spurious channels.

The sarcolemmal sodium channel can be solubilized in deoxycholate and similar ionic detergents, but the channel is much less stable than in the nonionic detergents. A purification scheme has been developed that utilizes the zwitterionic detergent CHAPS; channel protein purified with this detergent appears comparable on gel electrophoresis to that obtained with NP-40, but specific saxitoxin binding activity is not well preserved and resultant specific activities are not as high as with our usual purification procedure. Presumably some sodium channel protein is denatured or undergoes an unfavorable conformational change during the purification. Approaches to improving the stability of the channel after solubilization in CHAPS are now being explored.

In spite of these problems we have observed single-channel activity in purified channel prepared in CHAPS and reconstituted into phosphatidylserine/phosphatidylethanolamine vesicles. Several classes of channel activity have been seen repeatedly in inside-out patches pulled from liposomes enlarged by freezing and thawing [34].

One class of channels appears to be voltage- and neurotoxin-insensitive and may represent channels damaged during purification. Generally these are seen as random channel openings of 1–50-msec duration with a conductance of ~ 15 pS in symmetrical solutions containing 100 mM NaCl at 22°C. Some of these channels appear to have a subconductance state of 8–10 pS. Other channels with the same 15-pS conductance alternate between low and high noise levels during the channel-open interval. Occasionally, a conductance event of 30–40 pS is seen in these patches.

A second and more interesting class of channel events shows voltage-dependent behavior in the presence of batrachotoxin. These channels typically exhibit only irregular bursts of openings in the absence of batrachotoxin but convert to continuous opening flickers within seconds of application of the toxin. With 100 mM NaCl on both sides of the membrane, these toxin-activated channels demonstrate voltage-dependent activation (Fig. 3). At large hyperpolarizing voltages the channel is mostly closed, showing occasional discrete opening events. With progressive depolarization, channel openings become more frequent and the open time increases. At more depolarized voltages the channel is mostly open with brief transitions to the closed state. The voltage dependence of channel opening appears shallower than that seen with the channel in situ. It remains to be determined whether this difference reflects alterations in the channel or changes in the membrane environment itself.

Freeze-thaw appears to reorient the channels such that the tetrodotoxin binding site is on the outside of the vesicles, resulting in an "inside-out" orientation of patches on the electrode. Because of the resultant inaccessibility of the tetrodotoxin binding site after the patch is formed, it has not yet been possible to document TTX sensitivity in the channels that have shown this voltage-dependent activation.

Preliminary results with the patch-clamped, reconstituted channel are encouraging. Our experience suggests that the major limitation is purifying the channel in the presence of acceptable detergents without producing channel damage rather than in the technical aspects of single channel measurements with the reconstituted protein.

CONCLUSIONS

The voltage-sensitive sodium channel controls a complex relationship between membrane potential, time, and ionic currents. It contains binding sites for a variety of neurotoxins, some of which are allosterically coupled. Kinetic schemes for channel activity based on biophysical measurements suggests the presence of multiple intermediate states, presumably reflecting different channel configurations. In spite of the obvious complexity of channel function, the purified protein may be composed in its most primitve form of a single very large subunit. In higher animals one or more smaller subunits may be present, but the binding sites for tetrodotoxin and for the polypeptide neurotoxins still are found on the large subunit [34,35]. Unraveling the mechanisms through which a single polypeptide chain carries out the multiple complex and interrelated functions identified with the sodium channel will be a major task.

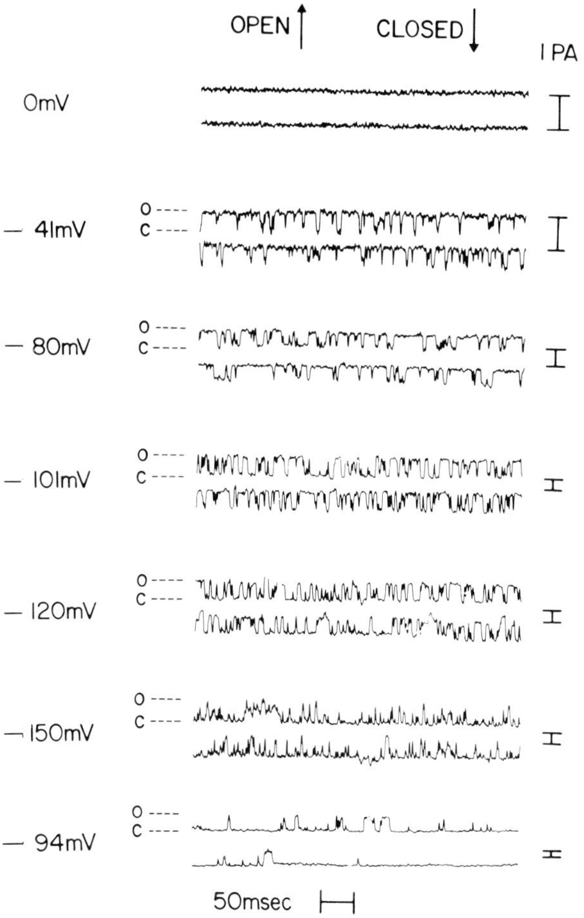

Fig. 3. Voltage-sensitive single-channel conductance events after batrachotoxin activation, measured using patch-clamp techniques, with purified rat sodium channels reconstituted into phosphatidyl serine/ phosphatidyl ethanolamine (PS/PE) vesicles. Batrachotoxin was used to remove channel inactivation in order to facilitate detection of channels; this toxin also shifts the voltage-dependence of channel activation toward more hyperpolarized potentials [30]. Single-channel events with a conductance of 15–18 pS are seen. The channel is mostly closed at hyperpolarizing potentials; the frequency and duration of channel openings increases with depolarization. At depolarized potentials, the channel is mostly open with brief transitions to the closed state. These recordings were made with 100 mM NaCl on both sides of the patch. The channel was oriented as in an inside-out cellular patch. The sign of the potential across the patch is given in terms of the normal cell membrane potential.

The purified mammalian sodium channel proteins retain most of their normal functional properties. They gate ionic fluxes in response to pharmacologic activators, and effectively select among monovalent cations. Preliminary data suggest that they also contain the machinery for voltage-dependent activation. These findings support the impression that the purified protein does indeed represent the entire channel rather than a single isolated component. Comparison of channel proteins from different sources underscores their strong similarities and suggests a high degree of conservation of protein structure during evolution. The major challenge for future work on the sodium channel will be to correlate emerging structural information on the channel protein with various aspects of its complex time and voltage dependent function.

ACKNOWLEDGMENTS

The authors' work described in this review was supported in part by N.I.H. grants NS-08075 and NS-18013, and by a grant from the Muscular Dystrophy Association. J.C. Tanaka is a postdoctoral fellow of the Muscular Dystrophy Association.

REFERENCES

1. Hodgkin A, Huxley A, Katz B: J Physiol 116:424–448, 1952.
2. Adrian RH, Chandler WK, Hodgkin AL: J Physiol 208:607–644, 1970.
3. Hodgkin A, Huxley A: J. Physiol 117:500–544, 1952.
4. Armstrong CM: Q Rev Biophys 7:179–210, 1975.
5. Agnew WS, Levinson SR, Brabson JS, Raftery MA: Proc Natl Acad Sci USA 75:2606–2610, 1978.
6. Barchi RL, Cohen SA, Murphy LE: Proc Natl Acad Sci USA 77:1306–1310, 1980.
7. Hartshorne RP, Catterall WA: Proc Natl Acad Sci USA 78:4620–4624, 1981.
8. Miller J, Agnew W, Levinson S: Biochemistry 22:462–470, 1983.
9. Barchi R: J. Neurochem 40:1377–1385, 1983.
10. Hartshorne R, Catterall W: J Biol Chem 259:1667–1675, 1984.
11. Weigele JB, Barchi RL: Proc Natl Acad Sci USA 79:3651–3655, 1982.
12. Tanaka JC, Eccleston JF, Barchi RL: J Biol Chem 258:7519–7526, 1983.
13. Tamkun MM, Talvenheimo JA, Catterall WA: J Biol Chem 259:1676–1688, 1984.
14. Barchi RL: Int Rev Neurobiol 23:69–101, 1982.
15. Catterall WA: Science 223:653–661, 1984.
16. Ritchie J, Rogart R: Rev Physiol Biochem Pharmacol 79:2–45, 1977.
17. Caterall WA: Annu Rev Pharmacol Toxicol 20:15–43, 1980.
18. Tamkun M, Catterall W: J Biol Chem 256:11, 457–463, 1981.
19. Kraner S, Tanaka J, Matesik D, Barchi R: (submitted for publication, 1984).
20. Agnew W, Raftery M: Biochemistry 18:1912–1919, 1979.
21. Cohen SA, Barchi RL: Biochim Biophys Acta 645:253–261, 1981.
22. Barchi RL, Murphy LE: J Neurochem 36:2097–2100, 1981.
23. Hartshorne R, Coppersmith J, Catterall W: J. Biol Chem 255:10,572–10, 575.
24. Norman R, Schmid A, Lombet A, Barhanin J, Lazdunski M: Proc Natl Acad Sci USA 80:4164–4168, 1983.
25. Haimovich B, Bonilla E, Casadei J, Barchi R: J Neurosci 4:2259–2268, 1984.
26. Casadei J, Gordon R, Lampson L, Schotland D, Barchi R: Proc Natl Acad Sci USA 81:6227–6231.
27. Barchi R, Tanaka J: Biophys J 45:35–37, 1984.
28. Frelin C, Vigne P, Lazdunski M: Eur J Biochem 119:437–442, 1981.
29. Huang LM, Catterall WA, Ehrenstein F: J Gen Physiol 73:839–854, 1979.
30. Huang LY, Moran N, Ehrenstein J: Proc Natl Acad Sci USA 79:2082–2085, 1982.
31. Quandt FM, Narahashi T: Proc Natl Acad Sci USA 79:6732–6736, 1982.
32. Yoshii M, Narahashi T: Biophysical J 45:184a, 1984.
33. Patlak J, Horn R: J Gen Physiol 79:333–351, 1982.
34. Lombet A, Norman R, Lazdusnki M: Biochem Biophys Res Commun 114:126–130, 1983.
35. Beneski DA, Catterall WA: Proc Natl Acad Sci USA 77:639–642, 1980.

Journal of Cellular Biochemistry 26:147–156 (1984)
Neurobiology: Molecular Biological Approaches to
Understanding Neuronal Function and Development 21–30

Detection of mRNAs Containing Regulatory Peptide Coding Sequences Using Synthetic Oligodeoxynucleotides

Illana Gozes, Mordechai Bodner, Yael Shani, and Mati Fridkin

Departments of Hormone Research (I.G., M.B., Y.S.) and Organic Chemistry (M.B., M.F.), The Weizmann Institute of Science, Rehovot, Israel

To understand the regulation of the production of peptide hormones, it is vital to elucidate their biosynthetic pathways. We chose to study a major regulatory peptide, vasoactive intestinal peptide (VIP), a peptide possessing both neurotransmitter and neurohormone actions. To identify the specific peptide mRNA we are using, as hybridization probes, radiolabeled synthetic oligodeoxynucleotides with sequence complementary to the predicted peptide mRNA sequence. Employing this approach, we identified and partially purified a ~1600-base long mRNA containing VIP related sequences which can be translated *in vitro* into VIP-immunoreactive polypeptides. Such mRNA was detected in normal VIP producing tissue (rat brain), as well as in a tumor producing VIP (human buccal tumor). This mRNA differs in size from a known VIP-mRNA identified in human neuroblastoma cells, suggesting the possibility of different VIP-mRNAs in different cell types.

Key words: VIP, oligodeoxynucleotides, mRNAs

Regulatory peptides exhibiting multiple functions may be controlled both at their production site and their function loci, ie, receptor site. We chose to study one such molecule, vasoactive intestinal peptide (VIP), a peptide with a broad biological activity, including induction of prolactin release through the hypothalamus-pituitary axis [1]. This peptide is widely distributed throughout the body in nerve cells, as well as in endocrine cells [2].

Hormonal activity of VIP is mediated through cyclic AMP [1]. Evidence supporting the neurotransmitter function attributed to VIP, in addition to its presence in neurons, include its localization in synaptosomal preparation [3] and its potassium induced release from those preparations [3,4]. VIP may also act as a neuromodulator. Indeed, the peptide and acetylcholine are simultaneously released from the parasympathetic nerve that innervates the submandibular salivary gland [5]. Acetylcholine is known to act via muscarinic receptors in activating salivary secretion and VIP potentiates this action. Recent experiments by Lundberg and colleagues [6] have

Received April 1, 1984; revised and accepted June 15, 1984.

shown that the affinity of muscarinic receptors for acetylcholine is increased by 10^5-fold in the presence of VIP. This may be the basis for the potentiating effect of VIP on cholinergic response. VIP is also involved in systemic vasodilation, smooth muscle relaxation [7], and glycogenolysis in the periphery [8] and in the brain [9].

In view of the variety and the multitude bioactivities of the peptide, it is plausable that several controlling mechanisms exist for its generation. To understand the regulation of VIP production, it is essential to elucidate its biosynthetic pathways. Peptide hormones are usually produced in minute quantities and the identification of their respective mRNAs might thus be thought as an unattainable task. However, current methodology enables the chemical synthesis of specific oligodeoxynucleotides probes, with sequences deduced from the peptide amino sequence, which can thereafter be used to identify and characterize the appropriate mRNA. The nucleic acid probes can be of unique sequences, following the predictions and rules derived by Agarwal et al [10]. Alternatively, owing to the extensive redundancy of the genetic code, it might be advantageous to use all possible sequence combinations, in order to decrease mismatches in hybridization [11]. The recent developments in the production of synthetic oligodeoxynucleotides [10–12], along with the possibility of obtaining very high specific radioactively labeled probes [13] and the progress in RNA-DNA hybridization methodology [14], have greatly increased the feasibility of studying biosynthesis of regulatory peptides.

Using the above approach, we were able to detect and partially characterize a mRNA containing VIP-related sequence in a human buccal tumor [15]. We have now extended these studies, and by the use of a battery of oligodeoxynucleotide probes specific for VIP, we have identified this mRNA in rat brain as well. Thus, the same mRNA containing VIP sequence is expressed in normal, as well as in a tumorogenic tissue of different species.

MATERIALS AND METHODS

Materials

Protected nucleosides: 5′Dimethoxy trityl thymidine, 5′dimethoxy trityl N-benzoyl adenosine, 5′dimethoxy trityl N-benzoyl cytidine, and 5′dimethoxy trityl N-isobutyryl guanosine were purchased from Vega Biochemicals (Tucson, AZ). Rats used were of the inbred Wistar strain obtained from an in house colony. Oligo(dT)-cellulose was purchased from Collaborative Research (Waltham, MA). Radiochemicals and enzymes were obtained from Amersham Radiochemical Center (Amersham International plc., England).

Synthesis of Oligodeoxynucleotides

Methyl phosphorochloroamidite was reacted with each of the protected nucleosides essentially as described [12]. The chemically reactive nucleoside dimethylamino phosphoramidites obtained were used as the building blocks in the oligodeoxynucleotide synthesis. The synthesis was conducted in a stepwise manner on a modified silica as a solid support [12]. The sequence of the synthetic chain was confirmed by the Maxam and Gilbert approach [16]. For radioactive ^{32}P labeling, the procedure described by Richardson [13] was employed. In short, 20 pmol of oligodeoxynucleotide were labeled with 20 pmol γ-labeled ^{32}P-ATP (7,000 Ci/mmole) in the presence of 8 U T_4 polynucleotide kinase. The probes were separated from the free ^{32}P-ATP by gel filtration on a Sephadex G-50 column.

Preparation of Poly(A)-Rich mRNA

Brain RNA was prepared from mature rat brains (Wistar rats, 4–8 weeks old) according to the procedure described before [17]. Briefly, frozen tissues (1 volume) were homogenized in the presence of 9 volumes phenol, m-cresol (8/1; vol/vol) and 9 volumes of 4% Na-para-amino salicylic acid in H_2O and RNA was extracted into the aqueous phase. Tumor RNA was prepared from HEp 3 (a neoplasm derived from a lymph node containing metastatic epidermoid cancer, primary in the buccal mucosa [18], which can be passed in nude mice). The tumor was kindly provided as frozen tissue by Professor Ruth Miskin and Mr. Jonathan Axelrod. For tumor mRNA preparation, the urea-LiCl [19] method was adapted. RNA precipitated in 6 M urea containing 3 M LiCl was then extracted with phenol-chloroform mixtures [19]. Poly(A)-rich RNA was obtained by oligo(dT)cellulose affinity chromatography [20].

mRNA Blot Analysis

Poly(A)-rich mRNA was subjected to electrophoresis on 1.5% agarose slab gels in the presence of 3.25% formaldehyde and transferred to nitrocellulose filters [21,22]. Hybridization to ^{32}P-labeled oligodeoxynucleotide probes was performed as described [15].

Separation of RNA on Sucrose Gradients

RNA was separated on sucrose gradients essentially as described previously [15]. For VIP mRNA localization, samples were spotted onto nitrocellulose filters which were presoaked in 10 × SSC (which is 10 times concentrated SSC and 1× SSC = 0.15M NaCl, 0.015 M Na citrate) and then processed for hybridization as above [15].

Translation of VIP Putative mRNAs In Vitro

Reticulocyte lysate was purchased from Amersham Radiochemical Center, and mRNA was translated according to the protocol provided by the company. ^{3}H-leucine was used to label the newly synthesized proteins. The translation reaction was carried out in the presence of human placental RNAse inhibitor (1 U/μl), purchased from Amersham. The putative peptide precursors were immunoprecipitated by the addition of rabbit anti-VIP antiserum (Antibody No. 5600-9, [23], a gift from Dr. J. Fahrenkrug, Clinical Chemistry Department, Frederiksborg Amts Centrabyglehus, 3400 Hillerod Sweden). Immunoprecipitation was performed according to Obata et al [24]. $NaDodSO_4$-polyacrylamide gel electrophoresis and autoradiography were performed as described previously [17,25].

RESULTS

A mRNA Containing VIP Sequences in Rat Brain

In this paper, we show that a peptide-related mRNA can be identified using mixtures of synthetic oligodeoxynucleotides, of lengths $\geqslant$17 bases, with sequences predicted from the peptide's amino acid sequence. This technique may be generally applicable to the study of the biosynthesis of regulatory peptides. Our preliminary studies with a short 11-base long VIP oligodeoxynucleotide probe did not give a specific hybridization to a single mRNA band (I. Gozes, R.J. Milner, R.A. Lerner, and F.E. Bloom, unpublished results). The DNA probe contained the unique nucleotide sequence: 3′…TTT.GTT.TAC.CG…5′, corresponding to the VIP sequence, starting from amino acid 15 as follows: .lys.glu.met.ala…[2]; this probe was custom synthesized by Collaborative Research (Waltham, MA). We concluded that, in the case of VIP, a probe with a unique sequence of 11 bases may be too short for specific hybridization and may recognize irrelevant mRNAs. In addition, the specificity of hybridization is not determined only by the length of the probe. Thus, when longer oligodeoxynucleotides of mixed sequences are used, an increased number of sequences present can obscure any specific hybridization. For example, a mixture of 17-base long probes for VIP prepared for us by Biologicals (Toronto, Canada) through McNeil Pharmaceuticals (Spring House, PA) (containing 64 different sequences that are actually a mixture of probes 1-4 described below) showed hybridization to quite a few RNA bands (I. Gozes, R.J. Milner, R.A. Lerner, and F.E. Bloom, unpublished results). However, the possibility of multiple mRNA species containing VIP sequences cannot be ruled out.

For further studies and precise characterization of mRNAs containing regulatory peptide sequences, we are using an efficient system for facilitated synthesis of oligodeoxynucleotides and we use these compounds to identify and partially characterize the specific mRNAs.

We synthesized eight different oligodeoxynucleotides (four specific for the N-terminal region of the VIP molecule and four specific for the C-terminal region of the VIP molecule) as hybridization probes for the identification of VIP-mRNA, as outlined below.

The VIP sequences is as follows [2]:

5 10 15 20 25
HisSerAspAlaValPheThrAspAsnTyrThrArgLeuArgLysGlnMetAlaVal LysLysTyrLeuAsn SerIleLeuAsnNH_2

We chose two stretches of amino acids of limited codon redundacy, N-terminal-related probes—amino acids 6–11 and C-terminal related probes—amino acids 15–21.

N-terminal probes

6
Peptide: . . .Phe.Thr.Asp.Asn.Tyr.Thr. . .

mRNA: 5′..$\mathrm{UU^{U}_{C}}$.ACN.$\mathrm{GA^{U}_{C}}$.$\mathrm{AA^{U}_{C}}$.$\mathrm{UA^{U}_{C}}$.AC. . . . .3′

Probes: 1 3′. .$\mathrm{AA^{A}_{G}}$.TGT.$\mathrm{CT^{A}_{G}}$.$\mathrm{TT^{A}_{G}}$.$\mathrm{AT^{A}_{G}}$.TG. .5′
2 —A—
3 —C—
4 —G—

(Each of these probes is a mixture of 16 different unique sequence probes).

C-terminal Probes

15
Peptide: . . .Lys.Gln.Met.Ala.Val.Lys.Lys. . .

mRNA: 5′.$\mathrm{AA^{A}_{G}}$.$\mathrm{CA^{A}_{G}}$.AUG.GCN.GUN.$\mathrm{AA^{A}_{G}}$.AA. . 3′

Probes: 5 3′.TT$^{T}_{C}$.GT$^{T}_{C}$.TAC.CGT.CAN.TT$^{T}_{C}$.TT. . 5′
6 —A—
7 —C—
8 —G—

(N denotes any nucleotide: each of the above probes is a mixture of 32 unique sequence probes).

The oligodeoxynucleotides described above were hybridized to mRNA from a known VIP-containing tissue, ie, rat brain. The probes were first end-labeled with ^{32}P, as described in Materials and Methods, and thereafter hybridized to the mRNA which had been resolved on denaturing agarose gels. All eight probes were tested for hybridization at various temperatures (21°C–40°C), according to Dalbadie-McFarland [11]. Probe 1 showed specific hybridization to a 1,600–1,700-base mRNA, which was maintained up to 40°C (Fig. 1).

Probe No. 4 also showed hybridization to the same region, but in addition it hybridized to other mRNAs as well (one mRNA of about 1,300 bases and one of about 2,100 bases; the latter could possibly be 18S ribosomal RNA). At elevated temperatures, the hybridization to the additional (1,300 and 2,100 bases) mRNAs decreased relative to the hybridization to the 1,600–1,700 base band (at 34°C the hybridization to this band amounted to 45% of the total, while at 30°C it amounted to 28%) (Fig. 1). Thus, a single base change (which is the difference between probe 1 and 4), in this case, still allowed hybridization to a 1,600–1,700-base RNA band; however, this hybridization decreased at 40°C in contrast with the hybridization to probe 1 (please see above). Similarly, but less definitive, hybridization to both the 1,600–1,700-base mRNA and the ~1300-base mRNA was obtained with probe 2; the latter is in agreement with the known sequence of VIP complementary DNA from human neuroblastoma [26].

Of the C-terminal probes, the most substantial hybridization to the 1,600–1,700-base band was obtained with probe 6 as depicted in Figure 2. However, hybridization to additional bands was observed as well. Several washes with 2× SSC at increasing temperatures (up to 40°C) were performed in order to differentiate the specific hybridization to the various bands, resulting in no observable change in their relative intensity. This result could possibly be explained by the multiplicity of sequences in probe 6 (32 sequences) as compared to probe 1 (16 sequences) and the possible multiplicity of mRNAs' containing VIP related sequences. Indeed, probe 6 can recognize the ~1300-base VIP-related mRNA as well (Fig. 2), as was found with probe 4 (Fig. 1). A similar mRNA (12–13S) coding for VIP was characterized before [26]. We have thus identified an additional, as yet unknown, putative VIP mRNA of 1,600-1,700 bases that hybridizes to both C-terminal and N-terminal VIP-specific oligodeoxynucleotide probes.

A Similar mRNA Containing VIP Sequences in Normal and Tumor Tissues

The N-terminal probe 1 was used to screen different tissues for VIP gene expression. Using this approach, we discovered the existence of elvated amounts of VIP-mRNA produced by a human buccal Tumor (Fig. 3). Since this tumor, which can be readily passaged in nude mice, provides a very rich source of total mRNA (yields at least ten times more material than brain per g tissue) and, as we have recently discovered contains elevated amounts of VIP [15], we chose to use this source of material for further studies on VIP gene expression.

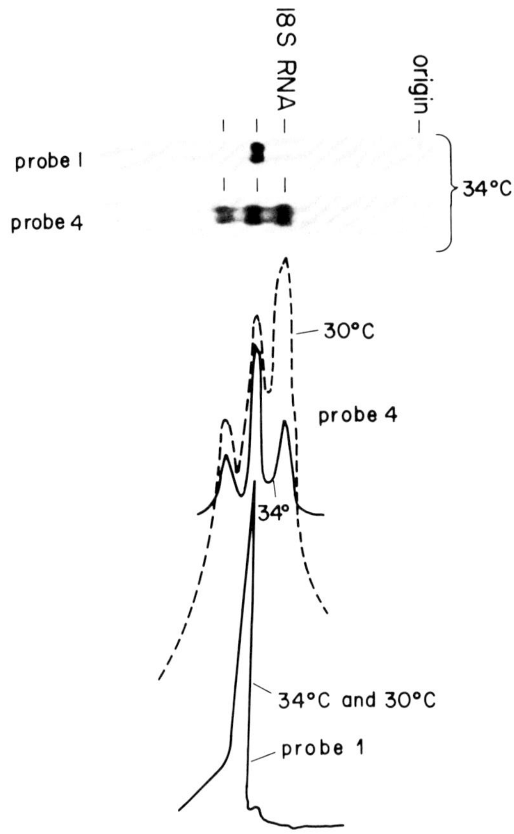

Fig. 1. Hybridization of synthetic oligodeoxynucleotides corresponding to the N-terminal region of VIP to brain mRNA. Brain mRNA prepared from mature rats [17] was resolved on agarose-formaldehyde gels and then transferred onto nitrocellulose filters [21,22]. Synthetic VIP specific oligodeoxynucleotides were end labeled with ^{32}P [13] and hybridized to this mRNA according to Stetler et al [14]. Following hybridization, the filters were exposed to autoradiography and the autoradiogram was subjected to densitometric scanning at 500 nm. The densitometric trances are depicted on the graph. It is worthwhile repeating here that one base alteration can at certain instances still allow hybridization. As depicted in the figure, oligodeoxynucleotide 4 (which has one base difference from oligodeoxynucleotides 1) also hybridized to the 1,600–1,700 bases mRNA, which is the only mRNA recognized by oligodeoxynucleotide 1.

We partially purified the putative VIP-mRNA from human buccal tumor (Fig. 3) using sucrose density gradient centrifugation (see Materials and Methods). The VIP-like mRNA fraction was thereafter identified using ^{32}P-labeled probe 1 as a VIP-specific hybridization probe [15]. The fractions giving the highest hybridization (binding the highest amount of radioactivity in dot hybridization) represented about 1,600-base long mRNA.

The final identification of the VIP-like mRNA species was achieved by its efficient translation into VIP-immunoreactive protein in a reticulocyte lysate cell-free system (Fig. 4). Further characterization of the translation products was obtained by immunoprecipitation, using specific anti-VIP antibodies. followed by gel electrophoresis and autoradiography. Various gradient fractions were processed in that way and

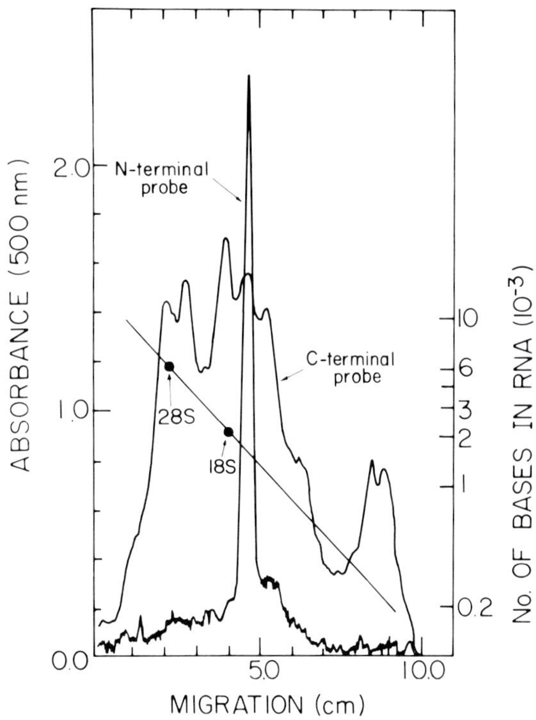

Fig. 2. Hybridization of synthetic oligodeoxynucleotide probes containing VIP coding sequences to brain mRNA. Experiments were performed as described in the legend to Figure 1. Oligodeoxynucleotide 1 (as a N-terminal hybridization probe) hybridized to a single mRNA band of 1,600–1,700 bases. Oligodeoxynucleotide 6 (as a C-terminal hybridization probe) hybridized specifically to the above RNA. Although this oligodeoxynucleotide hybridized to additional RNA bands as well, such hybridization could arise due to the multiplicity of sequences in probe 6 as compared to probe 1.

the ~ 1,600-base fraction gave distinct VIP-immunoreactive proteins, while other fractions did not.

An ~ 11,000-dalton protein band was identified using the antibody 5600-9. This is in close agreement with our previous observations using a different antiserum [15]. This protein may very well contain the entire VIP sequence since its binding to anti-VIP antiserum was totally inhibited by synthetic VIP [15]. It is particularly interesting to note that a protein having a similar molecular weight was proposed as the VIP precursor by pulse-labeling experiments in human neuroblastoma cells [27]. The 15,000-dalton VIP-related protein (Fig. 4) may contain VIP sequences as well. However, the major VIP-related protein, coded for by the mRNA containing VIP sequences, is an 11,000-dalton protein.

DISCUSSION

In this paper, we describe the use of synthetic oligodeoxynucleotides for the identification of mRNA species containing the VIP sequences. The possibility of multiple mRNAs coding for one peptide or for very similar peptides is demonstrated.

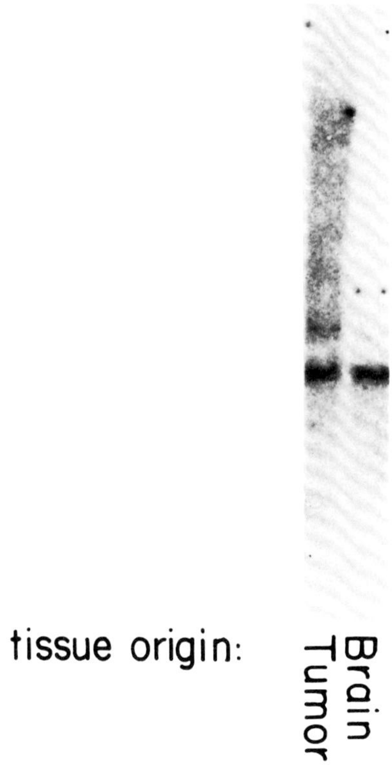

Fig. 3. Hybridization of VIP-related oligodeoxynucleotide No. 1 to tumor mRNA. Human buccal tumor mRNA and brain mRNA (prepared as described in Materials and Methods) were electrophoretically resolved and hybridized to probe 1 as in Figures 1 and 2. The picture shows the autoradiography. (Ribosomal RNA was used for molecular weight determination, see Fig. 2.)

Indeed, human neurobastoma prepro-VIP mRNA seems to be shorter than the 1,600–1,700-base mRNA that we have described (Figs. 1–3) and corresponds to ~1,300-base mRNA [26]. In addition, several high molecular weight proteins containing VIP-like sequences have been described, some of which are too high in molecular weight to be coded for by a 1,600–1,700-base mRNA [28–30]. Moreover, we identify here a noval source for VIP production namely a human buccal tumor [15]. This tumor also produces molecules similar to gonadotropin-releasing hormone (GnRH) and, thus, VIP and GnRH sequences may actually be found in the same cell (data not shown). The mRNA containing VIP-related sequences in brain, in the human buccal tumor and, as recently described, in a pancreatic tumor [31] are similar in size but differ from the one described for human neuroblastoma [26]. To further elucidate the VIP genomic organization, we used synthetic oligodeoxynucleotides specific for the peptide sequence and identified a recombinant phage inserted with human genomic DNA that is positive for the hormone sequence; elucidation of the gene structure is in progress (in preparation).

In conclusion, the resolving power of specific oligodeoxynucleotide hybridization may allow precise elucidation of regulatory-peptide biosynthesis and provide a

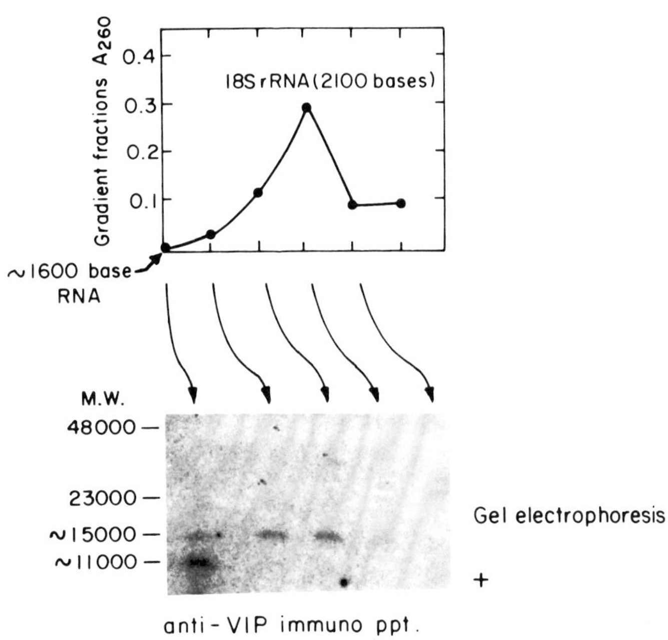

Fig. 4. Isolation of a mRNA containing VIP sequences on sucrose gradients and its in vitro translation. We isolated human buccal tumor mRNA and resolved it on sucrose gradients (see Materials and Methods). Samples were collected, spotted on nitrocellulose paper, and hybridized to ^{32}P-labeled oligodeoxynucleotide 1. The various samples were autoradiographed. The samples shown are of a size < 18S RNA, (see upper panel of RNA absorption profile at 260 nm). Samples > 18S did not show significant hybridization (Figs. 1–3). The gradient fractions were translated in a reticulocyte lysate cell-free translation system. The ^{3}H-leucine labeled newly synthesized proteins were immunoprecipitated with anti-VIP antibodies and subjected to $NaDodSO_4$ (15%) polyacrylamide gel electrophoresis (as described in Materials and Methods). The resulting autoradiograms of the gel are shown.

better understanding of multicellular interactions. Molecular cloning and rapid sequence analysis of the genes corresponding to the detected mRNA bands should yield information on sequences controlling gene expression and, in the long run, may shed light on the mechanism of cell differentiation and cell-cell communication.

ACKNOWLEDGMENTS

We are grateful to our collaborators from San Diego, Professors F.E. Bloom, R.A. Lerner, and Dr. R.J. Milner, for advice and support. We also wish to express gratitude to Professor R. Miskin and Mr. J. Axelrod for many helpful discussions and for the frozen human buccal tumor, and to Dr. J. Fahrenkrug for his kind gift of antibodies. We thank Mrs. Rona Levin for the typing and editing of this manuscript. This work was supported by grants (to I.G.) from NIH (NINCDS 1R1 NS 18960-01 A1), the United States-Israel Binational Science Foundation, the Fund for Basic Research administered by the Israeli Acadamy of Sciences and Humanities, the Jewish Agency, the Leo and Julia Forchheimer Center for Molecular Genetics, and by a Bergmann Memoral Prize to I.G. We are also grateful to Interpharm Inc. for supporting M.F. Dr. Illana Gozes is the incumbent of the Samuel O. Freedman Career

Development Chair in Cancer Research in perpetuity, established by the Montreal Chapter, Canadian Society for the Weizmann Institute of Science.

REFERENCES

1. Nicosia S, Oliva D, Giannathasio G, Spada A: J Endocrinol Invest 6:235, 1983.
2. Said SI: "Vasoactive Intestinal Peptide," Advances in Peptide Hormone Research. New York: Raven Press, 1982.
3. Giacchetti A, Said SI, Reynolds RC, Koniges FC: Proc Natl Acad Sci USA 74:3424, 1977.
4. Jessel TM, Iverson LL: Brain Res 143:174, 1978.
5. Lundberg JM: Acta Physiol Scand (Suppl 496): 112:1, 1981.
6. Lundberg JM, Hedlund B, Bartfai T: Nature 295:147, 1982.
7. Piper PJ, Said SI, Vane JR: Nature 225:1144, 1970.
8. Kerins C, Said SI, Proc Soc Exp Biol Med 142:1014, 1973.
9. Magistretti PJ, Morrison JH, Shoemaker WJ, Sapin V, Bloom FE: Proc Natl Acad Sci USA 78:6535, 1981.
10. Agarwal KL, Brunstedt J, Noyes BE: J Biol Chem 256:1023, 1981.
11. Dalbadie-McFarland G, Cohen LW, Riggs AD, Morin C, Itakura K, Richards JH: Proc Natl Acad Sci USA 79:6409, 1982.
12. Beaucage SL, Caruthers MH: Tetrahedron Lett 22:1859, 1981.
13. Richardson CC: Proc Natl Acad Sci USA 54:158, 1965.
14. Stetler D, Das H, Nunberg JH, Saiki R, Sheng-Dong R, Mullis KB, Weissman SM, Erlich HA: Proc Natl Acad Sci USA 79:5966, 1982.
15. Gozes I, Bodner M, Schwartz H, Shani Y, Fridkin M: Peptides 5:161, 1984.
16. Maxam A, Gilbert W: Proc Natl Acad Sci USA 74:560, 1977.
17. Gozes I, de Baetselier A, Littauer UZ: Eur J Biochem 103:13, 1980.
18. Moore AE, Sabachewsky L, Toolan HW: Cancer Res 15:598, 1955.
19. Auffray C, Nageotte R, Chambraud B, Rougeon F: Nucleic Acid Res 8:1231, 1980.
20. Aviv H, Leder P: Proc Natl Acad Sci USA 69:1408, 1972.
21. Nevins JR, Wilson MC: Nature 290:113, 1981.
22. Derman E, Krauter K, Walling L, Weinberger L, Ray M, Darnell JE Jr: Cell 23:731, 1981.
23. Fahrenkrug J, Schaffalitzky de Muckadell OB: J Lab Clin Med 89:1379, 1977.
24. Obata K-i, Itoh N, Okamato H, Yanaihara C, Yanaihara N, Suzuki T: FEBS Lett 136:123, 1981.
25. Gozes I, Schmitt H, Littauer UZ: Proc Natl Acad Sci USA 73:701, 1975.
26. Itoh N, Obatas K-i, Yanaihara N, Okamoto H: Nature 304:547, 1983.
27. Hioki Y, Hoshino M, Yanaihara C, Ogino K, Sato H, Yanaihara N: In Sakakibara S (ed): "Peptides Chemistry." Osaka: Protein Research Foundation, 1983, pp 263–266.
28. Gozes I, O'Connor DT, Bloom FE: Regul Peptides 6:111, 1983.
29. Gozes I, Milner RJ, Liu F-T, Johnson E, Battenberg ELF, Katz D, Bloom FE: J Neurochem 41:549, 1983.
30. Gozes I: In Zomzely-Neurath C, Walker WA (eds): "Gene Expression in Brain." John Wiley and Sons, Inc., (in press).
31. Delamarter J, Buell G, Kawashima E, Polak J, Bloom S: Regul Pept 7:283, 1983.

Journal of Cellular Biochemistry 27:181–187 (1985)
Neurobiology: Molecular Biological Approaches to Understanding Neuronal Function and Development 31–37

Domain Structure of Neurofilament Subunits as Revealed by Monoclonal Antibodies

Ruth Hogue Angeletti, John Q. Trojanowski, Martin Carden, William W. Schlaepfer, and Virginia M.-Y. Lee

Laboratory of Neuropathology, Department of Pathology and Laboratory Medicine, University of Pennsylvania School of Medicine, Philadelphia, Pennsylvania 19104

Monoclonal antibodies have been prepared against purified neurofilament (NF) subunits (NF68, NF150, and NF200). From 25 fusions, several hundred strongly positive antibodies have been obtained. Among them are antibodies against the specific subunits as well as antibodies recognizing common antigenic determinants. These have all been characterized according to the following properties: ELISA (enzyme-linked immunosorbant assay) testing against each subunit, immunoblots against enriched neurofilament preparation, immunoblots of cyanogen bromide or chymotrypsin-treated neurofilaments, immunofluorescence with PC12 cells, and immunohistochemistry of cerebellum. Whereas the antibodies against the NF68 and NF150 appear to react with single cyanogen bromide fragments, the antibodies against the NF200 react with multiple cyanogen bromide fragments. These data are consistent with the hypothesis that the NF200 is partially composed of several repeated structural determinants. Furthermore, all of the antibodies that react with the NF200 recognize the solubilized "sidearm" domain from limited chymotryptic digestions. The locations of the common and variable domains of the three subunits are discussed in light of these results.

Key words: neurofilaments, monoclonal antibodies, intermediate filaments

Neurofilaments are intermediate filaments of neurons and comprise the major cytoskeletal structure of large neuritic processes [1,2]. They are but one of several subclasses of intermediate filaments based on the size and differing immunoreactivities of protein subunits [3,4]. Most intermediate filaments are composed of a single major protein, which can be enriched by isolation and purification of the parent intermediate filament [5]. Neurofilaments, however, are unique in that they contain three major proteins with molecular weights of approximately 68,000 (NF68), 150,000 (NF150), and 200,000 (NF200) [6–8]. Although it has been suggested that these "triplet proteins" are structurally related, the evidence has been primarily indirect. In fact, peptide maps comparing the three polypeptides show no apparent similarities [9–12].

In contrast, immunological evidence has suggested that the three neurofilament polypeptides are related [13,14]. One monoclonal antibody has been developed that

Received April 17, 1984; revised and accepted September 24, 1984.

appears to detect a determinant common to all intermediate filaments [15]. Antibodies have recently been obtained that have reactivity against both subunit-specific epitopes or epitopes common to two or three of the subunits [16]. In the present manuscript, we report the analysis of the sets of neurofilament monoclonal antibodies that have been derived in our laboratory. In particular, it has been found that the NF200 appears to possess multiple repeated antigenic determinants. Furthermore, the distribution of common and unique epitopes within the polypeptide can be described.

MATERIALS AND METHODS

Bovine neurofilament proteins were prepared from spinal cord with some modification of the method previously described [17]. Monoclonal antibodies were prepared using electrophoretically purified neurofilament subunits as immunogens [17]. The hybrid clones were obtained from spleen cells of immunized Lewis rats fused with mouse myeloma cells. Positive clones were subcloned in agarose to obtain pure and stable monoclonal antibody-producing cells. Cyanogen bromide fragments (CNBr) were produced by incubation of purified neurofilament subunits with excess CNBr in 70% formic acid [18]. Soluble chymotrypsin fragments were obtained after limited digestion of enriched neurofilaments with chymotrypsin [5,19–21]. After digestion, the neurofilament preparation was centrifuged removing the undigested filaments and the low-molecular-weight "core" material from the reaction mixture. Analyses of intact and cleaved neurofilament samples were performed in sodium dodecyl sulfate on 6% or 7.5% polyacrylamide slab gels (0.75 mm) according to Laemmli [22]. The protein was transferred to nitrocellulose paper with an adaptation of the method of Towbin et al [23] utilizing a horizontal assembly. The electrophoretic blots were blocked for 1 hr in 1% normal rabbit serum and 1% bovine serum albumin in phosphate-buffered saline. Strips were then incubated overnight with the monoclonal antibodies diluted in 1% bovine serum albumin. After washing, the blots were developed with 4-chloro-1-naphthol.

RESULTS

Specificity of Monoclonal Antibodies

The hybridoma supernatants were tested for subunit specificity first by the ELISA test and then by immunoblotting enriched neurofilament fractions. Several thousand clones obtained from 25 fusions were screened, and only the strongest were maintained in culture. Typical results using the immunoblotting technique are seen in Figure 1. We obtained nine antibodies specific for the NF200, 25 specific for the NF150, and seven antibodies reactive only with the NF68. In addition, antibody-secreting clones were obtained that reacted with common determinants in two or more subunits. The largest of the cross reacting categories, 95 clones, was that identifying epitopes common to both NF150 and NF200. These included antibodies very strong or very weak against both polypeptides and antibodies that reacted strongly with one antigen and more weakly against the other. A single antibody was obtained that had specificity towards both NF68 and NF150; seven hybrids secreted antibodies reacting with NF68 alone. A single antibody was found that reacted with both NF68 and vimentin.

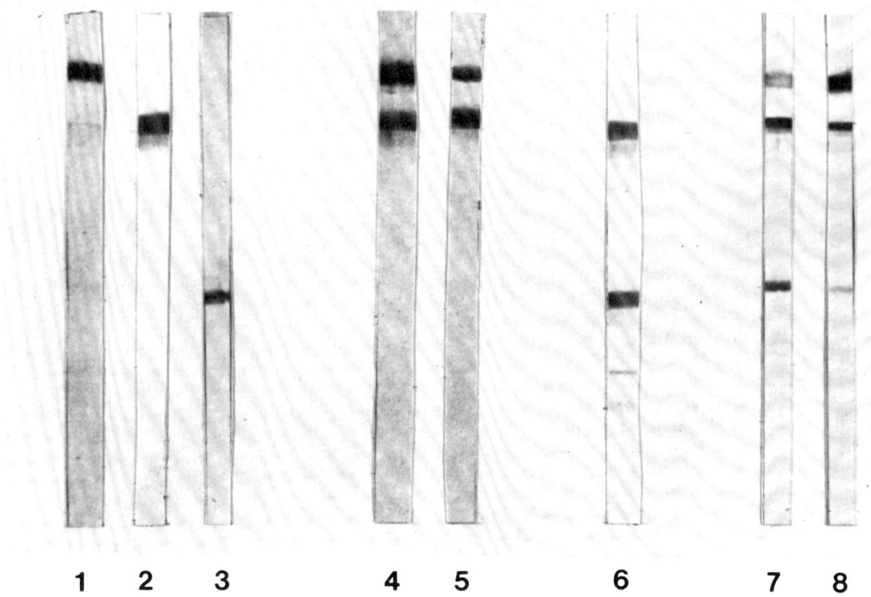

Fig. 1. Immunoblots of neurofilament preparation with selected monoclonal antibodies. Lane 1, anti-NF200; lane 2, anti-NF150; lane 3, anti-NF68; lanes 4 and 5, anti-NF150/200; lane 6, anti-NF68/150; lanes 7 and 8, anti-NF68/150/200.

Immunopeptide Mapping of Monoclonal Antibodies

In order to ascertain which portion of the individual neurofilament subunits reacted with specific antibodies, purified polypeptides were subjected to cyanogen bromide fragmentation, immunoblotted, and reacted with the hybridoma supernatants. Under these conditions, polyclonal antibodies usually react with multiple peptides, whereas monoclonal antibodies will react with unique peptides. In the case of the NF68 and NF150 immunopeptide maps, the latter case was most common (Fig. 2). However, using digests of the NF200, antibodies detected multiple peptide bands (Fig. 2). All antibodies that reacted with NF200 of whatever specificity (NF200, NF200/NF150, NF200/NF150/NF68) detected antigenic reactivity in the largest 125 kD fragment as well as in two smaller fragments. Subclones of these hybrid cells produced the same reactivity, indicating that true monoclonal antibodies were being produced.

The data obtained above for the NF200 indicate that this protein is formed in part from identical or closely related structural domains repeated throughout the polypeptide chain. Furthermore, the identity of this pattern of reactivity for NF200-reactive antibodies of all specificities indicates that several determinants are repeated in this manner.

Immunological Analysis of Chymotryptic Fragments

From work performed in several laboratories, it is known that short-term incubation of enriched neurofilament preparations with chymotrypsin produces sev-

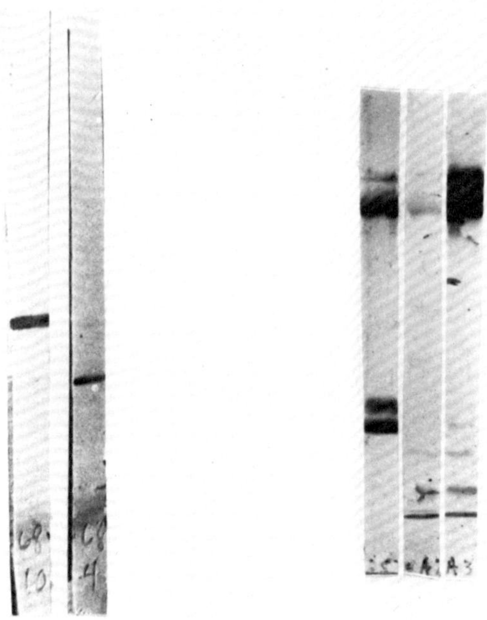

Fig. 2. Immunopeptide maps of neurofilament subunits. The individual neurofilament polypeptide chains were treated with cyanogen bromide and the resultant fragments separated on 7.5% (NF150) or 4–20% (NF68) sodium dodecyl sulfate polyacrylamide gels. After electrophoretic transfer, the nitrocellulose strips were incubated with monoclonal antibodies for the detection of immunoreactive fragments. On the left are two representative strips from anti-NF68 antibodies. On the right are three representative strips from anti-NF200 reactive antibodies whether subunit-specific or common determinants.

eral specific polypeptide fragments that are water soluble [19–21]. At the termination of the reaction, the filaments are centrifuged, removing unreacted neurofilaments plus a "core," α-helical region of approximately 40,000 daltons, presumably common to all neurofilaments. Indeed, it is this portion of the structure that is believed to be shared among all intermediate filaments [5,19–21]. The resultant supernatant contains four soluble polypeptides. A 160,000 dalton fragment is derived from the NF200, and a triplet of fragments at 130,000, 125,000, and 110,000 daltons is derived from the NF150 (Fig. 3). These soluble fragments, the carboxyterminal regions of the neurofilaments, are believed to represent "sidearms" from the main filamentous structure that extend exposed in the cytoplasm [24]. The structural differences between the neurofilament subunits are thought to reside in this region of the molecule (see model, Fig. 3). Our library of monoclonal antibodies was tested against chymotryptic digests of both assembled neurofilaments and purified subunits in order to determine where the unique and common epitopes lie.

Figure 4 shows examples of the typical reactivity of our antibodies. All antibodies that react with NF200, of whatever specificity, react with the solubilized 160 kD fragment. In turn, all of the NF150-specific antibodies react with the lower triplet polypeptides. In contrast, of the other antibodies that cross react with NF150, including antibodies that react with NF150/NF200 or NF68/150/200, only approximately one-third detect these solubilized fragments. The remainder also do not react

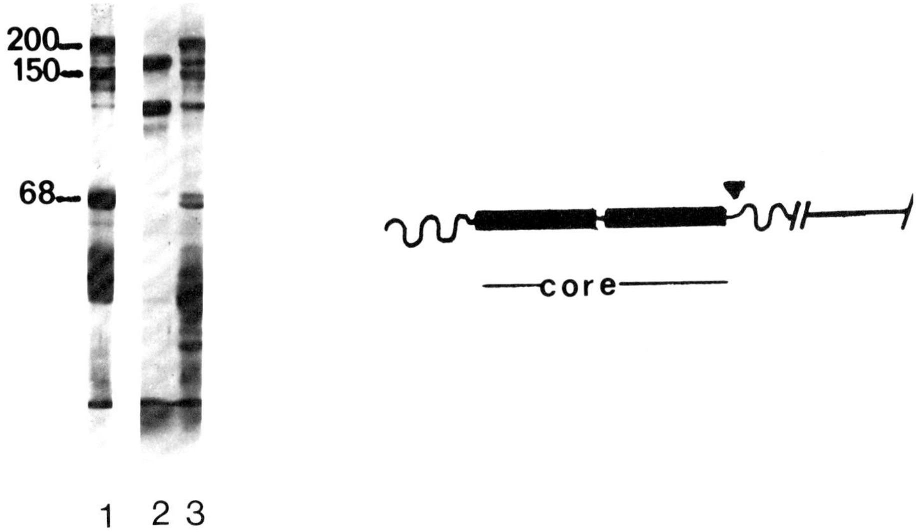

Fig. 3. Chymotrypsin digestion of intact neurofilaments. Left, sodium dodecyl sulfate electrophoretograms. Lane 1, enriched S1 neurofilament preparation; lane 2, supernatant from a 15′ digestion of S1 neurofilaments with chymotrypsin; lane 3, pellet from a 15′ digestion of S1 neurofilaments with chymotrypsin. Right, model of an individual neurofilament. A short amino terminal region is followed by an α-helical (two helices), or "core," region of ca 40,000 daltons. The carboxy-terminal region released by chymotrypsin (at arrow) is the location of the fragments released (lane 2 above). The "core" is thought to be a common element of all intermediate filaments [5,19–21].

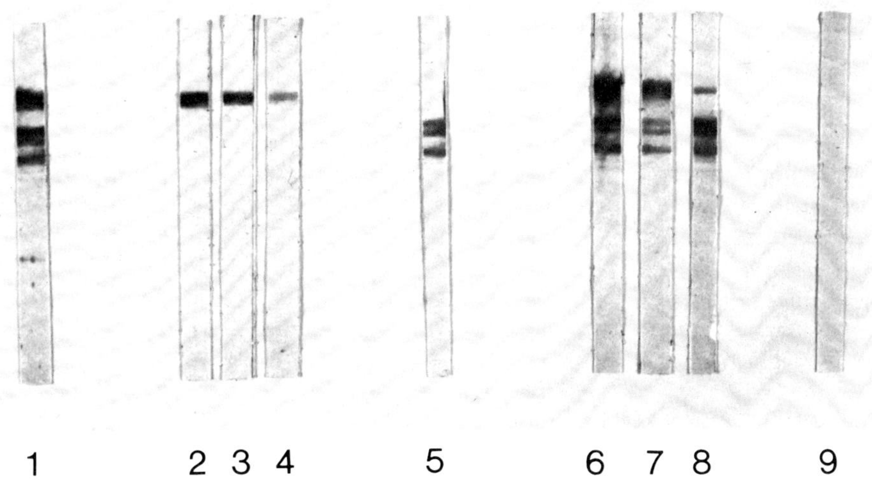

Fig. 4. Immunoblots of soluble chymotryptic fragments. Lane 1, polyclonal antiserum; lanes 2–4, typical reactions of antibodies that react with NF200; lane 5, reaction of antibodies specific to NF150; lanes 6–8, reaction of monoclonal antibodies that reveal both the fragments derived from NF200 and from NF150; lane 9, controls or negatives (see text for details).

with the "core," or common, region. Thus, it seems likely that the epitopes for these antibodies lie within the region of the polypeptide chain in which chymotrypsin cleaves. In addition, three of the NF68/NF150/NF200 antibodies react with the fragment derived from NF200, and a fourth antibody reacts with all of the solubilized fragments. None of these antibodies react with the "core" or presumably common region. Indeed, only five antibodies were found that did react with this insoluble "core" region. Three of these possessed reactivity unique to the NF68, one was specific to NF150, and the last cross reacted with both NF68 and NF150.

DISCUSSION

From these straightforward immunological peptide mapping experiments of neurofilaments we have been able to draw some straightforward conclusions. First, NF200 appears to have a portion of its structure composed of repeated domains. However, these epitopes lie wholly within the portion of the polypeptide that can be solubilized with chymotrypsin. Second, there are common antigenic determinants that lie within the fragments of both the NF150 and NF200 solubilized by chymotrypsin from intact neurofilaments. These regions of the molecules had previously been presumed to contain primarily variable determinants specific to the individual neurofilament subunits [5,20,21]. It should be mentioned that we do find subunit-specific determinants to be present in these regions as well. In addition, four of the antibodies that we have found to react with the "core" or putative common region instead react with determinants unique to specific subunits. Thus, the region of the molecule can contain variable epitopes as well as common ones. Finally, our antibodies that recognize all three neurofilament polypeptides are distinct from the monoclonal antibody previously found to react with all three as well as with all other intermediate filaments [15]. That antibody reacts with the "core" region; our antibodies react with the solubilized chymotryptic fragments.

In summary, our monoclonal antibodies have supported some of the hypotheses regarding the domain structure of intermediate filaments [5,20,21]. However, many of our antibodies reveal that the current model is too rigid and must be further elaborated. Molecular biological approaches currently in progress in several laboratories will be able to answer the questions raised. Specifically, from the above observations one would expect to find that the carboxy-terminal regions of the neurofilaments (sidearms) have several regions identical to one another. These identical sequences, plus the unique sequences, may have arisen by gene duplication events. These structural features are not reflected in the structures of the other intermediate filaments whose DNA sequences are already known [25–27]. Thus these features may be related to functions unique to the neurofilaments as well.

ACKNOWLEDGMENTS

This work was supported by grants from the National Institutes of Health NS-18362, NS-18161, and NS-15572 and by a grant from the March of Dimes, 1-826.

REFERENCES

1. Schlaepfer WW: Prog Neuropathol 4:101–123, 1979.
2. Norton WT, Goldman JE: In Bradshaw RA, Schneider DA (eds): "Proteins of the Nervous System." 1980, pp 301–309.
3. Lazarides E: Nature 283:249–256, 1980.
4. Lazarides E: Cell 23:649–650, 1981.
5. Steinert PM, Idler WW, Goldman RD: Proc Natl Acad Sci USA 77:4534–4538, 1980.
6. Schlaepfer WW, Freeman LA: J Cell Biol 78:653–662, 1978.
7. Liem RKH, Yen SH, Soloman GD, Shelanski M: J Cell Biol 79:637–645, 1978.
8. Anderton BH, Ayers M, Thorpe R: FEBS Lett 96:127–131, 1978.
9. Thorpe R, Delacorte A, Ayers M, Bullock C, Anderton BH: Biochem J 181:275–284, 1979.
10. Chiu FC, Korey B, Norton WT: J Neurochem 34:1149–1159, 1980.
11. Davison PF, Jones RN: Brain Res 182:470–473, 1980.
12. Brown BA, Nixon RA, Strocchi P, Marotta CA: J Neurochem 36:143–153, 1981.
13. Gown AB, Vogel AM: J Cell Biol 95:414–424, 1982.
14. Kjorell U, Thornell LE: Eur J Cell Biol 29:193–199, 1983.
15. Pruss RM, Mirsky R, Raff MC, Thorpe R, Dowding A, Anderton BH: Cell 27:419–428, 1981.
16. Lee VMY, Wu HL, Schlaepfer WW: Proc Natl Acad Sci USA 79:6089–6092, 1982.
17. Hogue-Angeletti R, Wu HL, Schlaepfer WW: J Neurochem 38:116–120, 1981.
18. Gross E: Meth Enzymol XI:238–255, 1967.
19. Julien P, Mushynski WE: J Biol Chem 258:4019–4025, 1983.
20. Geisler N, Kaufmann, Fischer S, Plessman U, Weber K: EMBO J 2:1295–1302, 1983.
21. Chiu TK, Eagles PAM, Maggs A: Biochem J 215:239–252, 1983.
22. Laemmli UK: Nature 227:680–685, 1970.
23. Towbin H. Staehelin T, Gordon J: Proc Natl Acad Sci USA 76:4350–4354.
24. Willard M, Simon C: J Cell Biol 89:198–205.
25. Hanukoglu I, Fuchs EV: Cell 33:915–924.
26. Quax-Jeuken YEFM, Quax WJ, Bloemendal H: Proc Natl Acad Sci USA 80:3548–3552.
27. Lewis SA, Balcarek JM, Krek V, Shelanski M, Cowan NJ: Proc Natl Acad Sci USA 81:2743–2746.

Journal of Cellular Biochemistry 27:189–203 (1985)
Neurobiology: Molecular Biological Approaches to
Understanding Neuronal Function and Development 39–53

How Do the Migratory and Adhesive Properties of the Neural Crest Govern Ganglia Formation in the Avian Peripheral Nervous System?

Jean Loup Duband, Gordon C. Tucker, Thomas J. Poole, Michel Vincent, Hirohiko Aoyama, and Jean Paul Thiery

Institut d'Embryologie du CNRS et du College de France, 94130 Nogent sur Marne, France

The peripheral nervous system derives mainly from the neural crest both in the head and trunk. Using markers such as fibronectin (FN), neural cell-adhesion molecule (NCAM), the nucleolar marker for quail cells in chimaeric embryos, and NC-1, a monoclonal antibody specific to crest cells and their neural derivatives, we have attempted to reconstruct the processes that lead to the formation of peripheral ganglia. Our observations allow us to propose a model of the formation of ganglia based on morphogenetic movements and on variations of crest cell adhesiveness. In most cases, crest cells migrate in morphologically defined and transient pathways that lead them to their final site of arrest; these pathways are always associated with FN, which appears necessary for crest cell attachment and movement in vitro. The directionality of crest cell migration is probably dictated by the cells' motile properties and population pressure in restricted areas suitable for cell movement. The disappearance of the pathways and of the substrate necessary for migration while the population is rapidly dividing may be responsible for the aggregation of crest cells in the case of the sensory ganglia. To the contrary, the aggregation of crest cells into autonomic ganglia (sympathetic, enteric, and ciliary ganglia) does not seem to obey the same rules, no disappearance of the substratum or of the pathways being obvious; rather, their formation seems correlated with the de novo synthesis of adhesive molecules such as NCAM.

Key words: avian embryo, peripheral nervous system, neural crest, cell-adhesion, cell migration

The peripheral nervous system arises entirely from the neural crest in the trunk and has a dual origin in the head (the neural crest and the ectodermal placodes for the sensory ganglia of some cranial nerves). Appropriate cell marking techniques have established very precisely the site of origin along the neural axis of the different ganglia [for an extensive review see 1].

Michel Vincent's present address is Département de Génétique Moléculaire et Humaine, Centre Hospitalier de l'Université Laval, Saint Foy, G1V 4G2 Canada.

Received April 1, 1984; revised and accepted July 27, 1984.

Let us, for example, consider the trunk level: Crest cells are able to differentiate into sensory and autonomic ganglia and into pigment cells. There is a striking contrast between the uniformity of the premigratory crest cells and the diversity of their differentiated derivatives. In addition, crest cells are able to migrate to widely dispersed areas of the embryo, even though at the onset of migration some of these areas do not yet exist. How can a single embryonic structure generate so many different ganglia in so many different places?

It is clear that the study of the initial and final states cannot solely explain this complex phenomenon. We have to consider carefully each of the intermediate states with special attention to the environment and its modifications. This raises several questions: Why do peripheral ganglia form in particular areas of the embryo? How do they form? What prevents crest cells from migrating in certain territories? Do the surrounding tissues control crest cell migration and, if so, how? What are the morphological cues that direct crest cell migration and arrest? How do the different cell lineages segregate?

Several hypotheses have been proposed to account partially for the precision with which crest cells reach their target sites. These include chemical repulsion from the neural tube or between crest cells themselves [2,3], contact inhibition of movement [4], contact guidance [5,6], haptotaxis [7], chemotaxis [8], and galvanotropism [9].

Here we examine the formation of most of the ganglia of the peripheral nervous system with a particular emphasis on morphogenetic processes and on adhesive properties of crest cells.

FORMATION OF THE SPINAL GANGLIA

The spinal or dorsal root ganglia (DRG) are present at levels posterior to the fifth somite. They lie along the side of the neural tube, among the sclerotomal cells that will differentiate into components of the vertebrae. This position proximate to the dorsal side of the neural tube suggested to early investigators that DRG derive from the neural crest [10]. Previous studies using neural crest markers proposed that crest cells invade the sclerotome but could not simply explain how and why they aggregate into sensory ganglia [7,11,12].

We have investigated this problem using markers of 1) extracellular matrix, 2) cell to cell adhesion, and 3) crest cells themselves. These markers are 1) fibronectin (FN), a glycoprotein that greatly enhances cell spreading, cell adhesion, and cell movement [for a review see 13]; 2) the neural cell-adhesion molecule (NCAM), an integral membrane glycoprotein that mediates the cell to cell adhesion in the nervous system [14]; and 3) the nucleolar marker for quail cells in chimaeric embryos [15] and NC-1, a monoclonal antibody that stains crest cells themselves and most of the neural tissues [16,17]. Using these tools, we have been able to reconstruct step by step the formation of most of the ganglia of the peripheral nervous system.

Soon after their dissociation from the neural epithelium, crest cells encounter an extracellular matrix between the ectoderm and the neural tube (Fig. 1a). This matrix is composed of a meshwork of fibers [4,6] of FN [18–21], collagen types I and III [22], and hyaluronic acid [23] and provides a substratum particularly suitable for migration (see below). Interestingly, this substratum lies between the basement membranes of the epithelia. The part of the basement membrane close to the epithe-

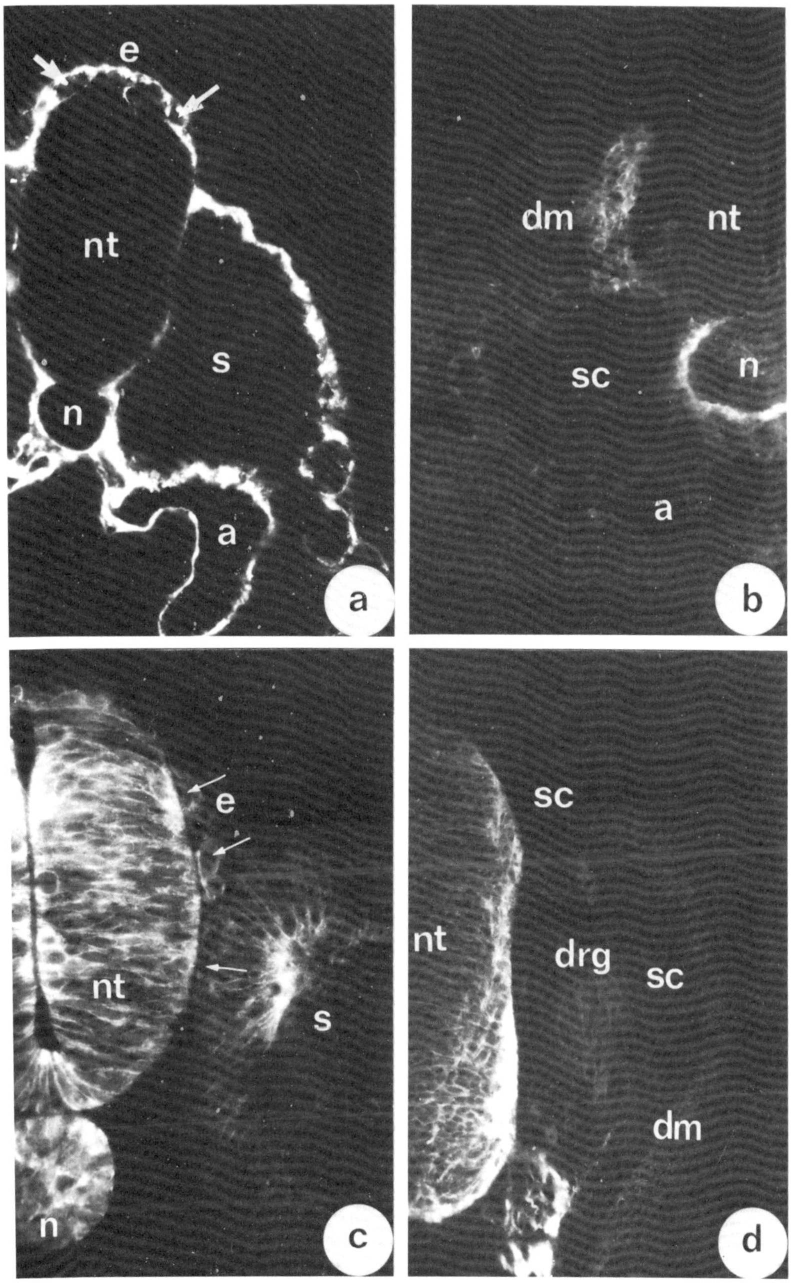

Fig. 1. Formation of the spinal ganglia. a) Immunolabeling for FN of an 18-somite embryo at the fifteenth somite level. FN is present in the basement membranes of the epithelia. Crest cells (arrow) start migrating in an FN-meshwork between the ectoderm and the neural tube. b) Immunolabeling of crest cells with NC-1 of a 32-somite embryo at the fifteenth somite level. Crest cells are only found in a pouch formed by the sclerotome, dermomyotome, and neural tube. c) Immunolabeling for NCAM of a 22-somite embryo at the fifteenth somite level. Though migrating, crest cells (arrows) still express NCAM on their surface. Note, however, that the pioneer crest cells are less stained, suggesting that they gradually lose NCAM. d) Immunolabeling for NCAM of an embryo at stage 23 of Hamburger and Hamilton [58]. The DRG rudiment is hardly stained as compared to the neural tube and motor nerves (compare with the staining of sympathetic ganglia at the same stage, Fig. 2d). a, aorta; dm, dermomyotome; drg, dorsal root ganglion; e, ectoderm; n, notochord; nt, neural tube; s, somite; sc, sclerotome.

lium, the basal lamina, contains laminin [22]. It is noteworthy that, as opposed to the organization of tissues in adults, such a configuration is predominant. Embryonic tissues are mainly organized in epithelial sheets, thus providing many potential pathways for crest cell migration.

After crest cells reach the somite, they begin to migrate as a dense population between the neural tube and the somite. While they progress ventrally, the somite starts to dissociate into the dermomyotome and sclerotome. The latter constitutes a dense mesenchyme that expands either by cell proliferation or by physical pressure [24] towards the notochord and neural tube. As a consequence, the potential pathway along the ventral side of the neural tube becomes obliterated. They accumulate along the lateral border of the neural tube (Fig. 1b), particularly at the level of the anterior part of the somite [our unpublished results], and progressively aggregate into a compact ganglion [see also 25]. Later on, sclerotomal cells will move dorsally and separate the DRG from the neural tube. It should be noted that some crest cells escape and move under the basement membrane of the dermomyotome [16,17]. These crest cells contribute to the Schwann cells of the sensory and motor roots [26, unpublished results].

Concomitant with the formation of the DRG, the extracellular matrix is considerably modified: FN and collagens disappear [21,22] and hyaluronic acid is partly replaced by chondroitin sulfate [23]. It has been proposed that this change in the glycosaminoglycan content [23] or that the disappearance of FN [21] could be responsible for the aggregation of crest cells. However, we cannot determine if the transformation of the matrix is a consequence or a cause of crest cell arrest.

It thus seems that the formation of the spinal ganglion is a consequence of the inability of crest cells to move further forward and backward due to the lack of available space and also of suitable substratum. In this respect, it was interesting to study the distribution of NCAM during the formation of the spinal ganglia. Prior to their migration, presumptive crest cells are associated with the neural epithelium and express NCAM on their surfaces. Surprisingly, the immunofluorescent staining for NCAM does not disappear from crest cells as soon as they detach from the neural tube (Fig. 1c). It should be noted however that the complete disappearance of NCAM from the surface of a cell is not required to get the complete dissociation of the cell from the others—a limited decrease of the number of NCAM molecules may be sufficient to induce a dramatic decrease in cell-cell adhesion [27]. As crest cells continue their migration towards the sclerotome, the staining for NCAM gradually disappears. During the accumulation of crest cells into DRG primordium, NCAM is not expressed significantly; its amount increases again at the time of neuronal differentiation (Fig. 1d), but it remains low compared to the neural tube, motor nerves, and autonomic ganglia (compare with Figs. 2d and 3e). Thus it seems that the compaction of crest cells into spinal ganglia is not rigorously correlated with the appearance of NCAM but rather to the disappearance of pathways of migration.

FORMATION OF THE SYMPATHETIC GANGLIA

The sympathetic ganglia are located in a mesenchyme under the sclerotome near the aorta. The grafting of tritiated thymidine-labeled neural tubes definitively confirmed the crest origin of these ganglia [11]. However, the way crest cells reach the aorta remained obscure and it was generally accepted that they penetrate through the sclerotome to become located to the ventral side of the embryo [7,11,28]. With this hypothesis, it was necessary to explain the segregation of the sensory and autonomic ganglia as they pass through the sclerotome.

We have investigated the possible presence of crest cells in the extracellular matrix of the intersomitic space. This space is filled partly by the intersomitic arteries derived from the aorta and by an FN-rich matrix between the somites (Fig. 2a). Crest cells rapidly invade this area and within a few hours reach the aorta as a continuous sheet (Fig. 2b,c). Since at this stage the aorta lies dorsal to the endoderm, crest cells cannot migrate further ventrally and spread in a rostral and caudal direction along the aorta. These cells will provide the sympathetic ganglia and the aortic plexuses. Moreover, the aorta is surrounded by an important meshwork of FN, which has been proposed as favoring the adrenergic differentiation of crest cells [29]. The intersomitic pathway is very transient due to the rapid expansion of the sclerotomes and their fusion into the vertebrae. Consequently, sympathetic precursors are precociously separated from the neural tube and the DRG rudiments.

Even though the final localization of crest cells along the aorta can be explained, it is difficult to understand the rapid aggregation of some of them into sympathetic ganglia, because, unlike the situation in the spinal ganglia, crest cells are not locked in a pouch that induces them to aggregate. Furthermore, the behavioral differences between sympathetic precursors aggregating rapidly and plexus precursors remaining for a long period as isolated cells cannot be explained from our data. Nevertheless, we have observed that some crest cells express NCAM very rapidly (Fig. 2d) and form small clusters in contrast to their neighbors, which spread along the aorta and whose NCAM expression is considerably delayed. The disappearance of the extracellular matrix cannot be responsible for the aggregation of crest cells, since the aortic region remains for a long time one of the areas richest in FN in the embryo. It should be noted however that FN and collagens become excluded from the ganglion cell mass as it differentiates [21,22]. Another as yet unrecognized matrix component might possibly influence the motility of crest cells. It seems rather that the aggregation of crest cells into sympathetic ganglia is a secondary event not directly related to their arrest. The de novo expression of NCAM by rapidly dividing cells is likely to be responsible for the formation of the sympathetic ganglia. In this respect, it is interesting to note that in vitro dissociated sympathetic neurons exhibit a strong tendency to move and reaggregate into small ganglia [unpublished results]. One should also remember that crest cells continue to proliferate actively while they already express neuronal markers such as the enzymes of catecholamine metabolism [30].

Finally, it is noteworthy that sympathetic cells undergo at 7 days a second migration, which leads them more laterally [31]. This phenomenon is not yet understood but it probably involved a mechanism different from that of the initial formation of the sympathetic ganglia, since it concerns mature neurons themselves.

The case of the superior cervical ganglia, the most rostral sympathetic ganglion, is much more complex. This ganglion can be considered as a polyganglion: It spreads along several somites and crest cells from the level of the first five somites contribute to its formation [32]. Moreover, regional differences in neural connectivity have been described in the superior cervical ganglion of both the rat [33] and the chick [34].

Neither the formation of this ganglion nor the pathways of migration of the crest cells from which it derives are yet well understood; it seems probable that it is

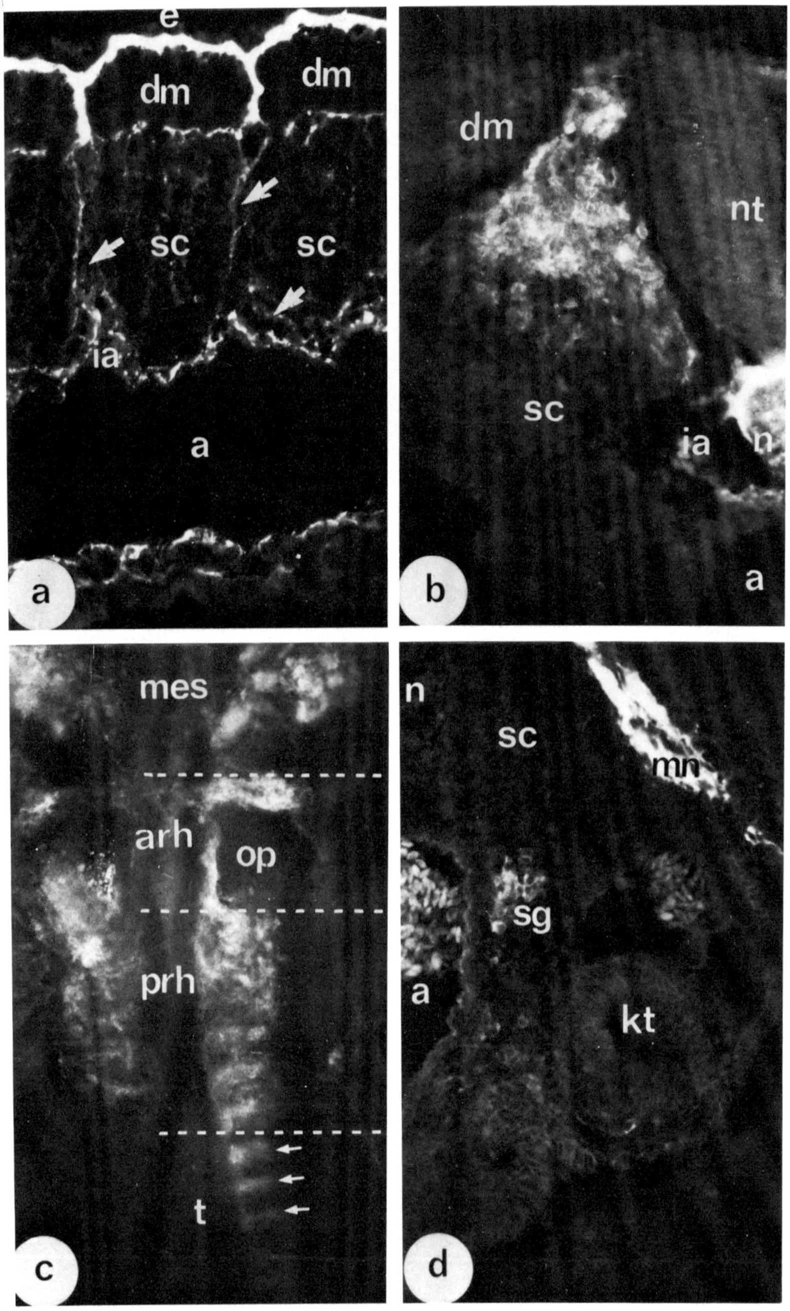

Fig. 2. Formation of the sympathetic ganglia. a) Sagital section of a 28-somite embryo stained for FN and showing crest cells in an FN meshwork along the intersomitic arteries. b) Transverse section of a 28-somite embryo stained with NC-1 at an intersomitic level. Crest cells occupy most of the area along the neural tube from the dorsal aspect of the embryo to the intersomitic arteries (compare with Fig. 1b). c) Whole mount of an 18-somite embryo stained with NC-1. The ectoderm has been peeled off over the head and on the right side of the trunk. At cephalic levels, crest cells migrate as a broad flux under the ectoderm except in the otic placode area. At trunk levels, crest cell fluxes are thin and located in a repeated manner between the somites (arrows). d) Transverse section of a stage 23 embryo stained with NCAM. The sympathetic primordium along the aorta is intensely stained. a, aorta; arh, anterior rhombencephalon; dm, dermomyotome; e, ectoderm; ia, intersomitic artery; kt, kidney tubule; mes, mesencephalon; mn, motor nerve; n, notochord; nt, neural tube; op, otic placode; prh, posterior rhombencephalon; sc, sclerotome; sg, sympathetic ganglion.

composed of initially independent sympathetic ganglia which fuse together during the large and complex morphogenetic movements in the head and neck (especially flexion and curvature). The study of its genesis is further complicated by the presence at this level of numerous ganglia and connective tissues deriving from the neural crest [32,35–38] particularly the enteric precursors. We do not yet know if the superior cervical ganglion shares common precursors with the enteric ganglia (see below) or if it forms by the same processes as the trunk sympathetic ganglia.

FORMATION OF THE PARASYMPATHETIC GANGLIA IN THE GUT

The enteric ganglia are distributed along the entire digestive tract. They are dispersed as a single or double ring of plexuses around the endoderm and among the mesenchymal cells that differentiate mainly into smooth muscles. Although found along the entire gut, the enteric ganglia originate from a precise region of the neural axis: from the vagal region extending from the first to the seventh somite and to a lesser extent from the caudal region [37,39] but not from the trunk region as was formerly thought [40,41]. The precise timing of the gut colonization and neuronal differentiation has been established using chimaeric embryos [39] and grafts of pieces of gut to the chorioallantoic membrane [42]. The colonization of the gut by crest cells is the longest migratory process in the embryo (from day 2 to day 9 of incubation). Here, again, the initial position of the crest cells and the final localization of the enteric ganglia raise questions regarding their formation. Many pathways and mechanisms might be proposed to account for their localization.

The initial steps of crest cell migration at the vagal level take place in an extracellular matrix between the basement membranes of the ectoderm and dermomyotome and between the sclerotome and the dermomyotome [38] leading them to the border of the somatosplanchnopleural epithelium (Fig. 3a). No extensive ventral migration was observed, correlating with the early dissociation of the somite into sclerotome and dermomyotome [21]. At the border of the somatopleure crest cells divide into two streams. One migrates between the ectoderm and the somatopleural epithelium and the other on the side of the splanchnopleural epithelium facing the endoderm. The enteric ganglia derive from the crest cells that take the latter pathway. These crest cells accumulate between the endoderm and the splanchnopleural epithelium. When the dorsal border of the splanchnopleural epithelium starts to delaminate into a thin epithelium and a loose mesenchyme, crest cells penetrate this mesenchyme and migrate caudally, remaining dorsal to the endoderm (Fig. 3b). Later on, crest cells may migrate ventrally as soon as the delayed delamination of the ventral border of the splanchnopleure occurs. During this invasion of the anterior part of the gut, crest cells mostly stay along the splanchnopleural epithelium as a one-cell-thick confluent layer.

In the foregut, the migration of crest cells approximately follows the closure of the gut. As they move, they are delayed by the formation of evaginations of the endoderm (hepatic, biliary, and pancreatic buds). Caudal to the yolk stalk, crest cells encounter the differentiating smooth muscle layer, which they do not seem able to invade. They are found on both sides of the muscle, trapped between the serosa and the circular muscle layer but rather dispersed and far less numerous between the muscle and the endoderm (Fig. 3d). The presence of this muscle layer might account for the formation of two concentric plexuses rather than a secondary migration through muscle cells toward the endoderm.

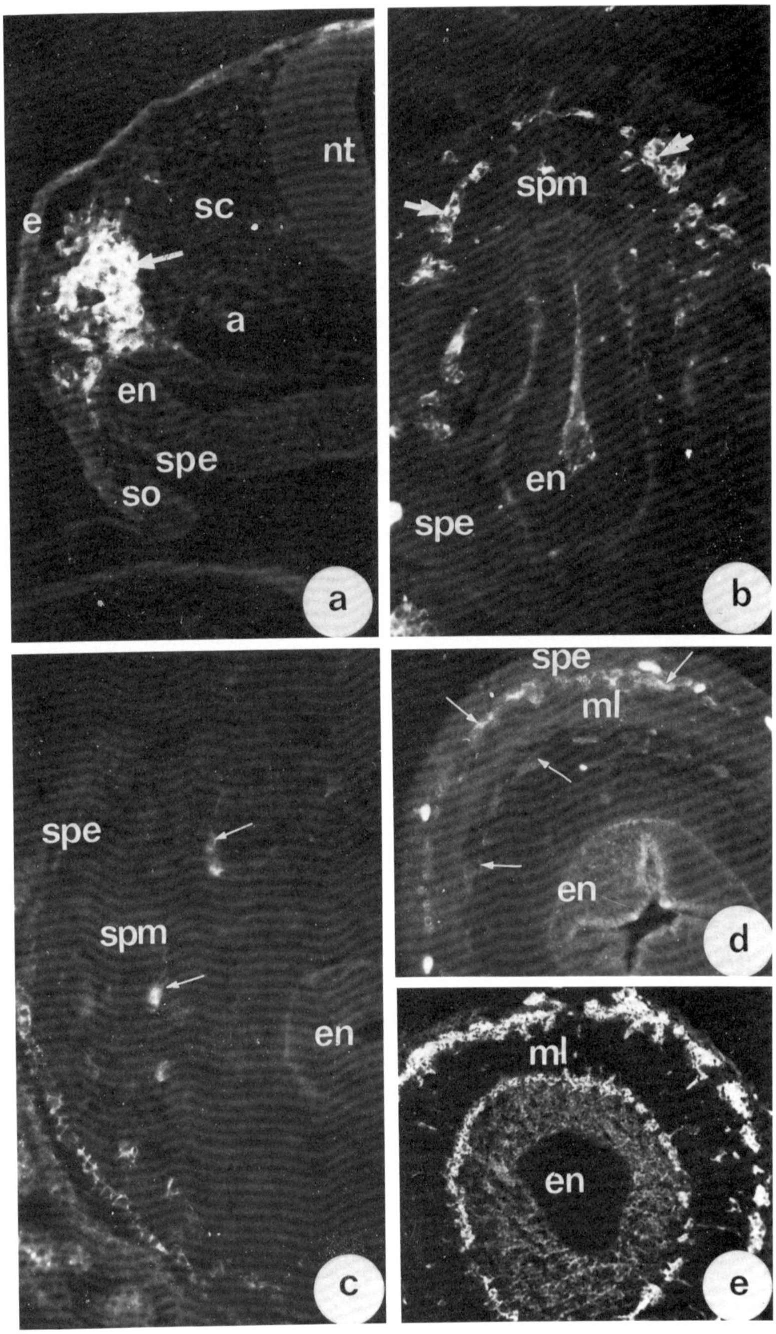

Fig. 3. Formation of the enteric ganglia. a) Migration of crest cells stained with NC-1 at the first somite level of a 22-somite embryo. Crest cells are accumulated on the lateral border of the endoderm. b) Staining of a 36-somite embryo at the level of the developing gizzard reveals neural crest cells (arrows) lining the sinuous border of the splanchnopleural epithelium in its delamination process. c) At 5 days of incubation, crest cells are NCAM positive and start growing axons (arrows) in the proventriculus. d) At 7 days of incubation, enteric precursors (arrows) have passed the umbilicus and the caeca. NC-1 staining shows precursors of the external plexus trapped between the developing longitudinal muscle layer and the well developed circular one. Crest cells belonging to the apparent internal edge of migration are less heavily stained. Note that they are found close to the circular layer and not along the side of the endoderm. e) 11-day-old embryo labelled for NCAM. The two plexuses in the rectum and their interconnecting fibers stain brightly. a, aorta; e, ectoderm; en, endoderm; ml, smooth muscle layer; nt, neural tube; so, somatopleure; spe, splanchnopleural epithelium; spm, splanchnopleural mesoderm.

As assessed by the presence of NCAM on their surface, crest cells aggregate immediately behind the leading edge of migration and they even start growing neurites (Fig. 3c,e). It therefore seems that the front of migration of crest cells divides intensely and is mainly responsible for the progression of crest cells. As was the case for the sympathetic ganglia, there appears no obvious reason for the crest cell arrest since the extracellular matrix does not disappear and no morphological barriers are interposed ahead of them. The appearance of NCAM might be due to the presence of mesenchymal components that induce a rapid differentiation of the crest cells and their arrest. The de novo synthesis of NCAM, concomitant with rapid cell proliferation, might further induce the formation of small ganglia.

We know very little about the pathways and the contribution of caudal crest cells to the innervation of the gut. However, their contribution to the intrinsic innervation of the gut is very limited [42,43]. They participate mostly in the formation of the ganglion of Remak located in the dorsal mesentery.

FORMATION OF THE SENSORY GANGLIA IN THE HEAD

The formation of the ganglia along the cranial nerves involves the neural crest and in some cases the ectodermal placodes. The use of radioisotopic labeling and of the quail chick chimaeras has indicated the relative contribution of the crest and placodes in the complex formation of these ganglia [35,36,44,45]. In all ganglia, the glial cells are derived from the neural crest. In some (geniculate, petrosal, and nodose), the neurons have a placodal origin; in others (jugular and superior), the crest gives rise to both the neuronal and glial populations. Finally, in the trigeminal ganglion both crest and placode cells form neurons.

Let us consider only the migration and aggregation aspect of the formation of these cranial ganglia. The pathways of migration are well documented for those ganglia located close to the neural tube.

The trigeminal ganglion derives from crest cells arising in the anterior rhombencephalon and migrating along the basement membrane of the neural tube. When they encounter the anterior cardinal vein and the cephalic mesenchyme, they accumulate (Fig. 4a) and rapidly aggregate into a ganglion rudiment (Fig. 4b).

The other sensory ganglia that appear in the hindbrain can be divided into two groups according to their final location: the proximal and the distal ones. They both are derived from crest cells migrating out of the posterior rhombencephalon caudal to the otic placodes. Though most of the crest cells at this level migrate laterally under the ectoderm, some of them remain close to the dorsal aspect of the neural tube (Fig. 5a) and differentiate into the proximal ganglia (jugular and superior ganglia). The distal ganglia, ie., the geniculate, petrosal, and nodose ganglia, appear laterally at the site of thickenings of the ectoderm (the ectodermal placodes) and derive from crest cells migrating between the ectoderm and dermomyotome (Fig. 5a). Finally, the outgrowing of the cranial nerves that link the distal ganglia to the proximal ones may utilize crest cells that have migrated under the dermomyotome for their directional cues (Fig. 5b; see also above).

Although the aggregation into the proximal ganglia and also into the tregeminal ganglion can be explained by the same morphological influences as the spinal ganglia (the cephalic mesenchyme and the cardinal vein playing the same role as the sclerotome and NCAM being expressed after the aggregation into ganglion), we do not yet

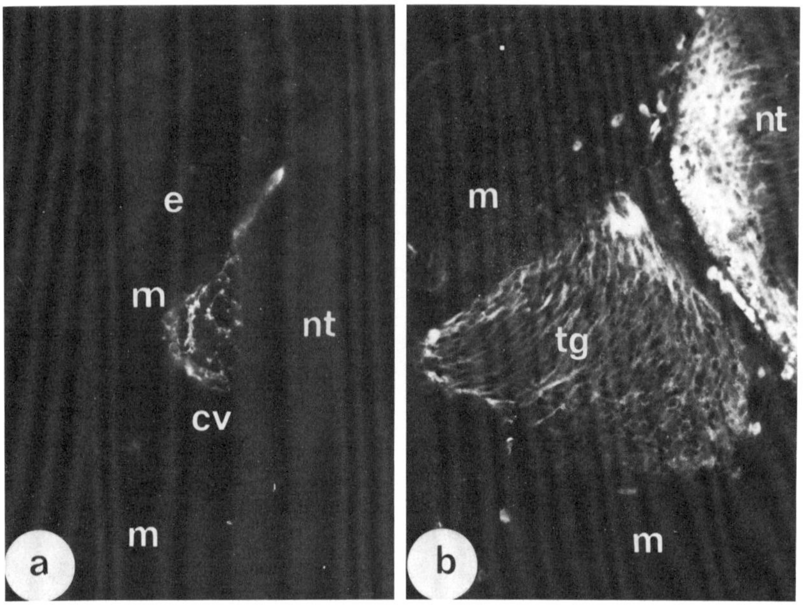

Fig. 4. Formation of the trigeminal ganglion. a) Staining with NC-1 of a 22-somite embryo at the anterior rhombencephalon level. Crest cells are trapped on the border of the anterior cardinal vein, between the neural tube and the cephalic mesenchyme. b) NCAM staining of a stage 21 embryo showing the well developed trigeminal ganglion among the cephalic mesenchyme. cv, cardinal vein; e, ectoderm; m, mesenchyme; nt, neural tube; tg, trigeminal ganglion.

understand how the distal ones form. Indeed the crest cells from which they derive are intermingled with mesenchymal precursors of the pharyngeal pouches, which are able to move further ventrally. Interestingly, the thickening ectodermal placodes express NCAM very early, along with some of the crest cells located very near to them (Fig. 5c,d). One might imagine that a mechanism of specific recognition via NCAM is responsible for the arrest of some crest cells and induces their integration into the placode.

FORMATION OF THE CILIARY GANGLION

Except for numerous autonomic plexuses (the autonomic paraganglia), the ciliary ganglion is the only autonomic ganglion in the head. It is located behind the optic vesicule near the optic nerve and innervates the muscles of the iris and the choroid. The crest cells from which it derives originate from the mesencephalon [46].

The mechanism of formation of this ganglion is unclear; its precursor cells are part of a dense crest population, which also will give rise to connective tissues of the face. During migration, the ciliary precursor cells may be segregated from the

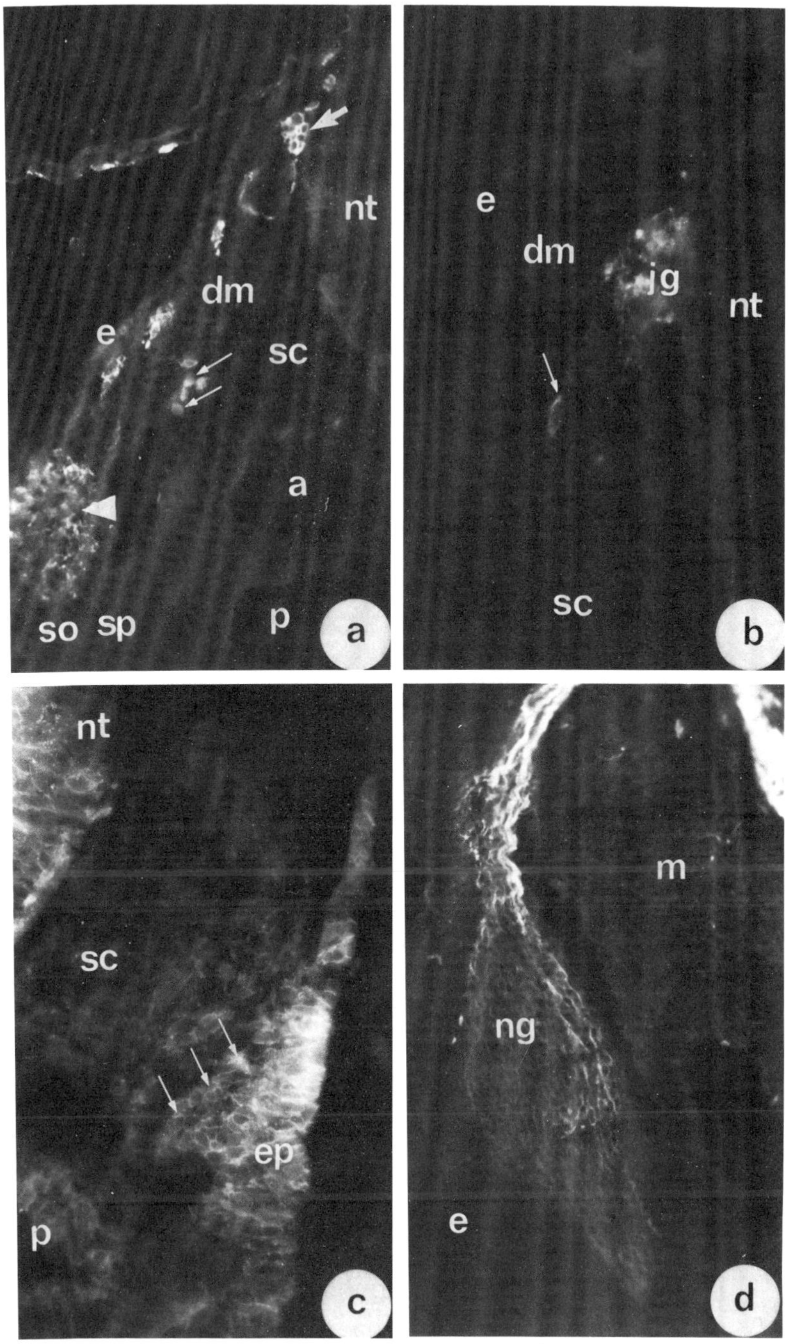

Fig. 5. Formation of the cephalic sensory ganglia. a) Immunolabeling for NC-1 of a 25-somite embryo at the level of the second somite. Crest cells are located mainly in three different places: above the sclerotome, trapped between the dermomyotome and the neural tube (short arrow), dispersed under the dermomyotome (thin arrows), and laterally along side the ectoderm. b) Immunolabeling for NC-1 of a 36-somite embryo at the level of the third somite. The developing jugular ganglion is apparent together with a glial cell accompanying a small neurite that expands from the ganglion. c) In a 28-somite stage embryo, NCAM stains the thickening ectodermal placode and numerous crest cells (arrows) close to it. d) At stage 21, the developed nodose ganglion is heavily stained for NCAM. a, aorta; dm, dermomyotome; e, ectoderm; ep, ectodermal placode; jg, jugular ganglion; m, mesenchyme; ng, nodose ganglion; nt, neural tube; p, pharynx; sc, sclerotome; so, somatopleure; sp, splanchnopleure.

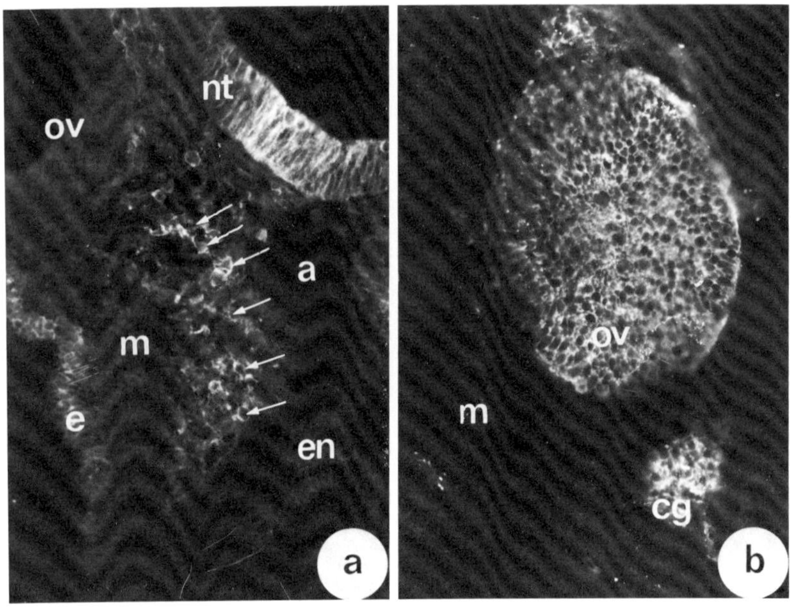

Fig. 6. Formation of the ciliary ganglion. a) A frontal section of a 22-somite embryo stained for NCAM reveals numerous stained cells along the aorta near the optic vesicule. b) NC-1 staining of a 25-somite embryo. Transverse section through the edge of the optic vesicule. The latter reacts strongly with the antibody and so does the developing ciliary ganglion. a, aorta; cg, ciliary ganglion; e, ectoderm; en, endoderm; m, cephalic mesenchyme; nt, neural tube; ov, optic vesicule.

mesectodermal cells both in vivo and in vitro [47–49], but no local segregation of these cell lineages has been detected except for the capacity of pioneer crest cells to produce FN. It has been suggested that crest cells synthesizing FN are the mesectodermal cell precursors that precociously express a mesectodermal property [18], but they may simply reflect a transient physiological property because migrating crest cells encounter a large space filled only with hyaluronic acid [50].

The routes of migration and the exact site of arrest of the ciliary ganglion precursor cells are not known. It has been proposed that these precursors migrate along growing nerves in the head [5]. However, using NCAM immunolabeling, we did not see any staining of growing neurites in the mesencephalon at the stage of crest cell migration. We have, however, detected some NC-1–positive cells exhibiting NCAM, which migrated as a dispersed population along the aorta towards the optic vesicule. These cells could be the ciliary precursor cells exhibiting early a neuronal marker (Fig. 6a,b). The mechanisms of arrest and aggregation of these cells into a ganglion are unknown but the morphology of the eye area suggests that they are similar to those of the other autonomic ganglia.

AN IN VITRO ASSAY FOR STUDYING CREST CELL MIGRATION AND AGGREGATION

We have developed in vitro cultures of crest cells that mimic the in vivo conditions of crest cell migration. When cultured on defined substrates, neural tube explants excised before crest cell emigration produce a halo of cells originating from the dorsal border of the neural tube. In the absence of FN, either in the medium or on the substratum, few crest cells emigrate from the neural tube. Collagen type I and laminin [51] along with hyaluronic acid [28] appear to be poor substrates for migration. Moreover, in vitro binding assays have shown that crest cells do not adhere efficiently to laminin and collagen. On the contrary, FN or FN-containing fibers deposited by fibroblasts are excellent substrates for crest cell adhesion and migration. When confronted with alternating stripes of FN and laminin or glass, crest cells preferentially migrate on FN whereas they aggregate on laminin and do not spread on glass. Finally, crest cells cultured for several days seemed to develop a higher affinity for laminin and progressively decreased their preference for FN.

Using time-lapse cinematography, we have found that the behaviors of crest cells differ according to their position relative to the entire population. Pioneer crest cells at the front of migration exhibit a random migration [2,3]. In contrast, crest cells located within the cell mass maintain a persistent direction of movement. The presence of surrounding cells in a limited environment apparently forces them to move in a straight line.

The role of FN has recently been defined through perturbation experiments. Monovalent antibodies to the cell-binding site of FN or peptides identical to sequence of the cell-binding domain when injected in crest pathways strongly inhibit crest cell displacement. These results demonstrate unequivocally that a direct interaction between the cell-binding site of FN and crest cells is an absolute requirement for their migration [52].

By varying the substrate of migration of crest cells, we obtained the formation of two- and three-dimensional clusters of cells. These clusters quickly exhibit NCAM. An in vitro microaggregation assay revealed that crest cells, according to their dissociation conditions, express either a calcium-dependant or a calcium-independant mechanism of aggregation. Furthermore, anti-NCAM monovalent antibodies strongly inhibit the calcium-independant aggregation of neural crest cells [53].

CONCLUSION

Our study demonstrates that the different ganglia of the peripheral nervous system are derived from crest cells that exhibit distinct patterns of migration and aggregation. However, the initial steps of migration share common properties characterized by several main features: 1) The substratum of migration is acellular and fibrous. Fibronectin, one of its main components, is necessary for transient cell adhesion and movement of crest cells; hyaluronic acid provides space, and collagens maintain the architecture of the tissues. 2) The pathways are organized as channels limited by the basement membranes of the epithelia and are dependent on the morphology of the tissues. 3) Crest cells move in these pathways as a dense cell mass, providing a unidirectional migration.

During the last steps of migration, the behavior of crest cells may vary according to the type of derivatives they give rise to. DRG and proximal ganglion precursors whose duration of migration is short keep a constant pattern of migration. On the contrary, autonomic precursors always begin migrating as a compact population and finish as a dispersed one. This is particularly true for the ciliary and enteric precursors, which migrate even in a cellular environment.

Furthermore, there is no strict correlation between the expression of NCAM and the migratory properties of autonomic precursors: NCAM is expressed by ciliary precursors while they migrate and enteric and sympathetic precursors exhibit NCAM concomitant with their aggregation.

The aggregation follows two types of mechanisms according to the fate of crest cells: 1) The spinal ganglia and the proximal ganglia in the head form as a consequence of the arrest of crest cells. This arrest is due to the presence of morphological barriers followed by the disappearance of substrate and available space for migration. Adhesive molecules are not involved in this process but appear at a low level on the differentiating neurons. 2) In the case of the autonomic ganglia and of the distal ganglia, NCAM appears to be required for crest cell aggregation because no obvious physical barriers are present.

Many aspects of crest migration remain to be elucidated. We cannot yet explain why crest cells sometimes do not utilize some pathways that exhibit, from our point of view, good criteria for cell migration; nor can we understand the mechanisms that trigger crest cell emigration from the neural tube.

Finally, our results, along with previous studies [54–57], suggest that some formerly proposed mechanisms such as chemical repulsion, chemotaxis, and haptotaxis are not likely to be involved in the migratory process of crest cells. However, we cannot exclude a possible local influence of contact inhibition of movement, contact guidance, or electric fields.

ACKNOWLEDGMENTS

We thank Drs. L. Blair and F. Dieterlen for critical reading of the manuscript and our colleagues Dr. R. A. Rovasio, A. Delouvée, and M. Denoyelle for fruitful collaboration.

This work was supported by grants from the Centre National de la Recherche Scientifique, the Institut National de la Santé et de la Recherche Médicale, the Ministére de l'Industrie et de la Recherche, and the Ligue Nationale Française Contre le Cancer.

H. A. was supported by a fellowship from the Institut National de la Santé et de la Recherche Médicale, the Society of Japan Promotion of Science, and the Fondation Fyssen; G. C. T. was supported by the Ecole Polytechnique and M. V. by NATO and the Natural Sciences and Engineering Research Council of Canada; T. J. P. was supported by the U.S.-France Exchange of Scientists Program of the National Science Foundation and the Centre National de la Recherche Scientifique and the Philippe Foundation.

REFERENCES

1. Le Douarin NM: "The Neural Crest," Cambridge: Cambridge University Press, 1982.
2. Twitty VC, Niu NC: J Exp Zool 108:405, 1948.
3. Twitty VC, Niu NC: J Exp Zool 125:542, 1954.

4. Tosney KW: Dev Biol 62:317, 1978.
5. Noden DM: Dev Biol 42:106, 1975.
6. Lofberg J, Ahlfors K, Fallstrom C: Dev Biol 75:148, 1980.
7. Weston JA: Adv Morphol 8:41, 1970.
8. Greenberg JH, Seppa S, Seppa H, Hewitt AT: Dev Biol 87:259, 1981.
9. Nucitelli R, Erickson CA: Exp Cell Res 147:195, 1983.
10. Tello JF: Trabajos Inst Cajal Invest Biol (Madrid) 39:1, 1947.
11. Weston JA: Dev Biol 6:279, 1963.
12. Noden DM: In Garrod DR (ed): "Specificity of Embryological Interactions," Receptor and Recognition, Series B, Vol. 4, London: Chapman and Hall, 1978, pp 1–49.
13. Hynes RO, Yamada K: J Cell Biol 95:369, 1982.
14. Edelman GM: Science 219:450, 1983.
15. Le Douarin NM: Dev Biol 30:217, 1973.
16. Vincent M, Duband JL, Thiery JP: Dev Brain Res 9:235, 1983.
17. Vincent M, Thiery JP: Dev Biol 103:468, 1984.
18. Newgreen DF, Thiery JP: Cell Tissue Res 211:269, 1980.
19. Mayer BW, Hay ED, Hynes RO: Dev Biol 82:267, 1981.
20. Duband JL, Thiery JP: Dev Biol 93:308, 1982.
21. Thiery JP, Duband JL, Delouvee A: Dev Biol 93:324, 1982.
22. Duband JL, Thiery JP: Cell Differ, submitted for publication.
23. Derby MA: Dev Biol 66:321, 1978.
24. Toole B: In Hay ED (ed): "Cell Biology of the Extracellular Matrix," New York: Plenum Press, 1981, p 259.
25. Pannese E: Adv Anat Cell Biol 47:1, 1974.
26. Teillet MA: personal communication.
27. Hoffman S, Edelman GM: Proc Natl Acad Sci USA 80:5762, 1983.
28. Newgreen DF, Gibbons IL, Sauter J, Wallensfeld B, Wutz R: Cell Tissue Res 221:521, 1982.
29. Sieber-Blum M, Sieber F, Yamada KM: Exp Cell Res 133:285, 1981.
30. Rothman TP, Gershon MD, Holtzer H: Dev Biol 65:322, 1978.
31. Kuntz A: J Comp Neurol 20:283, 1910.
32. Le Douarin NM: In Cowan M (ed): "Handbook of Physiology," in press.
33. Nja A, Purves D: J Physiol 264:565, 1977.
34. Yip JW: Abstr 13th Ann Meet Soc Neurosci, 1983, p 937.
35. Ayer-Lelievre CS, Le Douarin NM: Dev Biol 94:291, 1982.
36. D'Amico-Martel A, Noden DM: Am J Anat 166:445, 1983.
37. Le Douarin NM, Teillet MA: J Embryol Exp Morphol 30:31, 1973.
38. Tucker GC, Thiery JP: Dev Biol, submitted for publication.
39. Yntema CL, Hammond WS: J Comp Neurol 101:515, 1954.
40. Van Campenhout E: Physiol Zool 5:333, 1932.
41. Andrew A: J Anat 98:421, 1964.
42. Allan IJ, Newgreen DF: Am J Anat 149:413, 1977.
43. Teillet MA: Wilhelm Roux' Arch Dev Biol 184:251, 1978.
44. Noden DM: Dev Biol 67:296, 1978.
45. Noden DM: Dev Biol 67:313, 1978.
46. Narayanan CH, Narayanan Y: J Embryol Exp Morphol 47:137, 1978.
47. Le Douarin NM, Teillet MA: Dev Biol 41:162, 1974.
48. Barrald K: In Spitzer NC (ed): "Neuronal Development," New York: Plenum Press, 1982, p 101.
49. Ziller C, Dupin E, Brazeau P, Paulin D, Le Douarin NM: Cell 32:627, 1983.
50. Pratt RM, Larsen MA, Johnston MC: Dev Biol 44:298, 1975.
51. Rovasio RA, Delouvee A, Yamada KM, Timpl R, Thiery JP: J Cell Biol 96:462, 1983.
52. Boucaut JC, Darribere T, Poole TJ, Aoyama H, Yamada KM, Thiery JP: J Cell Biol, 99:1822, 1984.
53. Thiery JP, Tucker GC, Aoyama H: In Edelman GM, Gall WE (eds): "Molecular Basis of Neural Development," New York: John Wiley and Sons, in press.
54. Bronner-Fraser M, Cohen AM: Dev Biol 77:130, 1980.
55. Tosney KW: Dev Biol 89:13, 1982.
56. Erickson CA, Weston JA: J Embryol Exp Morphol 74:97, 1983.
57. Erickson CA, Olivier KR: Dev Biol 96:542, 1983.
58. Hamburger V, Hamilton HL: J Morphol 99:49, 1951.

Journal of Cellular Biochemistry 27:205–214 (1985)
Neurobiology: Molecular Biological Approaches to Understanding Neuronal Function and Development 55–64

The Differential Distribution of Beta Tubulin mRNAs in Individual Mammalian Brain Cells

W. Sue T. Griffin, Michael A. Alejos, Erica J. Cox, and Marcelle R. Morrison

Department of Cell Biology (W.S.T.G., M.A.A., E.J.C.) and Department of Neurology (M.R.M.), University of Texas Health Science Center, Dallas, Texas 75235

It has been shown by in vitro translation of polyadenylated messenger RNAs (poly(A)$^+$ mRNAs) that the mRNAs encoding both alpha and beta tubulin isotypes are present at much higher relative levels in the developing rat brain than they are in the adult [1], suggesting that the requirements for tubulin subunits vary with cell type and/or with the developmental stages of a particular cell type. The postnatally developing rat cerebellum, with its readily identifiable cell populations that perform the gamut of developmental tasks, is a suitable model for analyzing specific cellular mRNA distributions during development. In this report, by in situ hybridization techniques it is shown that, by comparison to total cellular poly(A)$^+$ mRNA levels, there is relatively more of the total beta tubulin mRNAs in mitotically active external granule layer cells than in those in the internal granule layer. These results show that migration and differentiation of these granule cells is accompanied by a decrease in their beta tubulin mRNA levels relative to the levels in granule cells of the external granule cell layer. Furthermore, the relative levels of beta tubulin mRNA both in the prenatally formed Purkinje cells and the postnatally formed stellate cells are two to fourfold less than in the granule cells of the internal granule cell layer.

Key words: tubulin, mRNA, in situ hybridization, rat cerebellum, development

Tubulin is required for several neuronal functions such as the assembly of microtubules for spindle formation [2,3], the growth and maintenance of neuronal processes [4], neurosecretion [4], the maintenance of cellular structure, and possibly migration [5]. Each of these functions is performed in one or another of the cell populations in the developing cerebellum of the 14-day-old rat [6,7]. Specifically, the granule cells of the matrix zone of the external granular layer (EGL) are dividing; the newly formed premigratory granule cells in the mantle zone of the EGL are extending axonal processes parallel to the pial surface and are perhaps "gearing up" for migration; granule cells migrating through the molecular layer toward the internal granular layer (IGL) are elaborating their axons; postmigratory basket and stellate

Received May 17, 1984; revised and accepted October 24, 1984.

cells are forming extensive dendritic arborizations that contact parallel axons of granule cells and axonal collaterals which terminate on dendrites and somata of Purkinje cells; the prenatally formed Purkinje cells are elaborating a vast dendritic network in the molecular layer; and the mature granule cells in the IGL are fully functional interneurons that are performing their differentiated tasks [5].

Our results [1,8] and those of others [9,10] show that the relative amounts of mRNAs encoding the major tubulin isotypes decrease during brain development. These results could be a consequence either of tubulin mRNA levels decreasing similarly in each cell type or of differential changes occurring in particular cell populations. The only way to differentiate between these possibilities is to use the technique of in situ hybridization to quantitate the relative levels of tubulin mRNAs in individual cell types at different stages of brain development.

In situ hybridization has been used successfully to localize the pituitary cells that synthesize proopiomelanocortin [11] and the pancreatic cells which synthesize preproelastase [12], as well as to study the induction of specific mRNAs during cell differentiation [13–16]. The specificity of this technique now has been documented in a number of systems [17–21], and several studies have shown it to be quantitative [13,18–22]. In this paper, the relative levels of the mRNAs encoding the beta tubulins are compared with total poly(A)$^+$ mRNAs in the various cell types of 14-day-old rat cerebellum. Plasmids containing chick beta tubulin DNA inserts [obtained from D. Cleveland, 23] were nick translated and the radiolabeled DNA probes were in situ hybridized to their complementary mRNAs within cells in sections of cerebella from 14-day-old rats. The distribution of the mRNAs was estimated by counting autoradiographic grains over individual cells of the various types and comparing the distribution of poly(A)$^+$ mRNA within each cell type by hybridization of [^{3}H]-polyuridylate (poly(U)) to adjacent tissue sections.

METHODS

Animals

In these experiments, 14-day-old Fischer rats were used.

Fixation and Embedding of Cerebellar Tissue

The vermis of the cerebellum was dissected free and fixed by immersion in Bouin's solution for 4 hr, then cleared with 70% ethanol. The tissue then was dehydrated through alcohols and xylenes and embedded in paraffin.

Sectioning and Slide Preparation

Sections were cut on a rotary microtome at 5 μm and collected on gelatin-chromalum subbed slides. Sections were deparaffinized in three changes of toluene and rehydrated through alcohols to water.

Preparation of Radioactively-Labeled DNA

Plasmid pBR322 with a beta tubulin insert and native pBR322 were radioactively labeled using a BRL nick translation kit and [^{3}H]-dCTP as the labeled nucleotide (59.2 Ci/mmole; NEN) to a specific activity of 6×10^5 cpm/μg and 3×10^6 cpm/μg DNA, respectively [24].

In Situ Hybridization of cDNA Probes

Either the nick-translated pBR322 probe or the beta tubulin probe was added to the hybridization buffer (10 mM Tris HCl, pH 7.6; 200 mM NaCl, and 5 mM $MgCl_2$), containing 500 μg/ml polyadenylate, 500 μg/ml sheared E coli DNA, and 25% formamide. The mixture was boiled 60 sec, quickly cooled, and then applied to the cerebellar sections (10 μl/slide; 6×10^4 cpm; 100 ng beta tubulin and 20 ng pBR322). The sections then were incubated in a moist chamber at 50°C for 1 hr, after which they were washed for 1 hr at 50°C in hybridization buffer, washed overnight in post-hybridization buffer (50 mM Tris HCl, pH 7.6, 10 mM KCl, 1 mM $MgCl_2$) at 4°C, dipped in distilled water, kept in ice cold 5% trichloracetic acid for 15 min, rinsed in distilled water and dehydrated through alcohols.

Autoradiography

Slides were dipped in NTB-2 Kodak emulsion and exposed for 5 days. Exposed slides were developed in Kodak D-19 and stained with hematoxylin and eosin y [25].

[^{3}H]-poly(U) Hybridization

Ten μl of hybridization buffer containing [^{3}H]-poly(U) (3 Ci/mmole, 4×10^4 cpm, 20 ng) was applied to cerebellar sections and incubated for 1 hr at 50°C in a moist chamber [13]. After hybridization, the slides were treated as described above for the tubulin probe hybridization. Before hybridization, control slides were incubated with RNase A (100 μg/ml) in RNase buffer (5 mM Tris HCl; pH 7.6; 1 mM KCl; 0.1 mM $MgCl_2$) for 1 hr at 37°C to degrade RNA, washed for 10 min under running tap water, and then were hybridized and treated as above [13,23].

Estimation of mRNA Content in Each Cell Type

Slides were examined at 400 to 1000 diameters and the autoradiographic grains per cell in microscopic fields, both over and under focused, were enumerated (grains above no less than 100 cells of each type were counted in cerebellar sections from 3 different animals in separate experiments). The average number of grains in each cell type in the pBR322 control sections were subtracted from the grains per cell in the sections hybridized to the beta tubulin DNA probe. Similarly, grains in the RNase-pretreated sections were subtracted from those in the sections which had been hybridized to poly(U).

Calculation of tubulin to poly(A)$^+$ mRNA ratios

The following equation was used to calculate the ratio of tubulin mRNA to total poly(A)$^+$ mRNA:

$$\frac{\text{Mean grains/cell type after tubulin cDNA hybridization}}{\text{Mean grains/cell type after poly(U) hybridization}}$$

RESULTS

In this study, the hypothesis was addressed that a cell type performing a developmental task contains relatively higher levels of specific mRNAs encoding proteins involved in that task. In particular, the distribution of beta tubulin mRNAs was determined relative to total poly(A)$^+$ mRNAs among the different cell types and stages of granule cell differentiation in the 14-day rat cerebellum.

We used in situ hybridization of [^{3}H] poly(U) to quantitate total poly(A) levels in each cell type [11,12,21,25,26]. Conditions of hybridization were such that the poly(U) was present in 100-fold excess. Each 5 μm section of 14-day rat cerebellum contains 5 ng of total poly(A)$^+$ mRNA, and assuming an average steady state poly(A) tail size of 80 nucleotides [27], each section contains 0.2 ng of poly(A). Therefore, 20 ng of [^{3}H] poly(U) were applied to each section. Similarly, the beta tubulin probe also was present in 100-fold excess, assuming beta tubulin mRNAs comprise 10% of total mRNAs in the 14-day cerebellum.

Figures 1 and 2A, B, and C show that the majority of grains for both the poly(U) and the beta tubulin probes were localized over cells. Few, if any, grains were seen over glomeruli in the IGL, as would be expected, as glomeruli contain axon terminals [7] and hence should contain no mRNA. Two controls were performed to show that hybridization was specific. First, poly(U) did not hybridize to sections pretreated with RNase A (Fig. 1D) and RNase T2 (results not shown) demonstrating that the poly(U) was indeed hybridizing to poly(A) regions in intracellular RNA. Second, few grains were seen over sections hybridized to a control pBR322 probe (Fig. 2D), demonstrating that the tubulin DNA insert was hybridizing specifically to complementary sequences in the sections shown in Figure 2A, B, and C.

Autoradiographic grains over EGL, IGL, stellate, and Purkinje cells were counted for both the poly(U) and tubulin hybridizations. Column 1 of Table I shows that the grain counts over each cell type were highly reproducible with a standard deviation of less than 10%. Moreover, each cell type had a distinct complement of both poly(A)$^+$ mRNAs and beta tubulin mRNAs. In neither case were grain yields optimized for by, eg, varying the time and temperature of hybridization. Thus, the percentage of total poly(A)$^+$ mRNA which is tubulin mRNA cannot be calculated in each cell type. However, others have shown that the ratio of grain yields in different cell types is independent of the absolute extent of hybridization [19,28], thus allowing us to compare relative poly(A)$^+$ and beta tubulin mRNA levels even if our hybridization conditions are not yet optimal.

It was found that the stellate cells had the highest level of poly(A)$^+$ mRNA, there being an average of 70.6 grains over these cells in 5 μm sections. Purkinje cells contained an average of 30.1 grains per cell and granule cells in the EGL contained 23.0 and in the IGL, 24.3 grains per cell. These results show that the relative amounts of poly(A)$^+$ mRNA in individual cell types is not directly proportional to cell size. For example, the average granule cell in a 5 μm section has approximately the same amount of poly(A)$^+$ mRNA as does the average Purkinje cell, although the average

Fig. 1. Autoradiographs of in situ hybridization of ^{3}H-poly(U). ^{3}H-poly(U) (3 Ci/mmole; 40,000 cpm/section) was hybridized to cerebellar sections from 14-day-old rats. A) Grain density over granule cells in the EGL. B) Grain density in Purkinje cells (P) as well as in migrating cells (M), stellate cells (S), and basket (B) cells in the molecular layer (MOL). C) Grain density in granule cells of the IGL. Note the paucity of grains over the mossy fiber axon terminals in the glomerulus (G). D) Grain density in Purkinje cells, migrating cells, stellate and basket cells, as well as in granule cells (g) of the IGL. The low density of grains over cells in D are due to pretreatment with RNAse A before hybridizations. Autoradiographic exposure was for five days. Magnification, ×500.

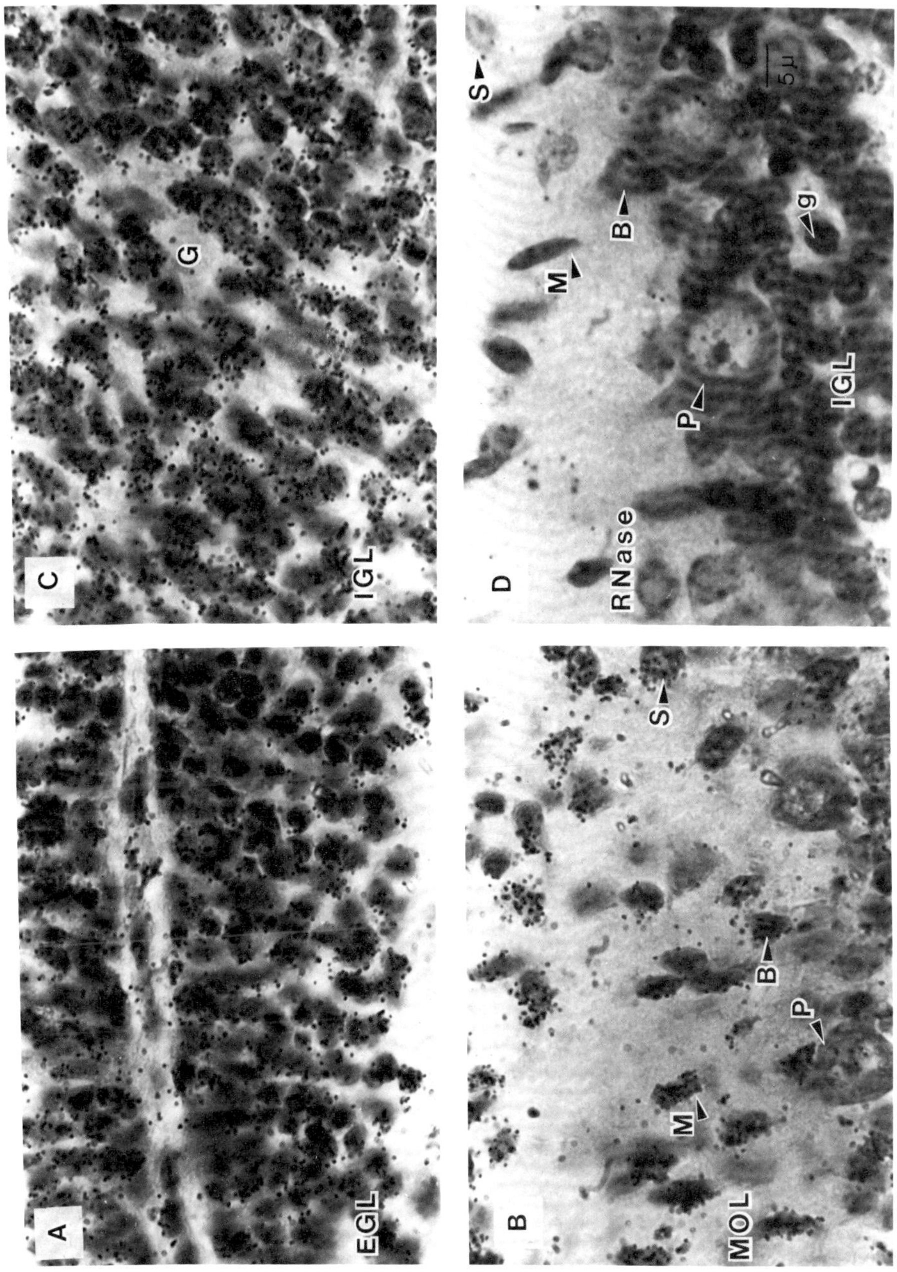

Fig. 1.

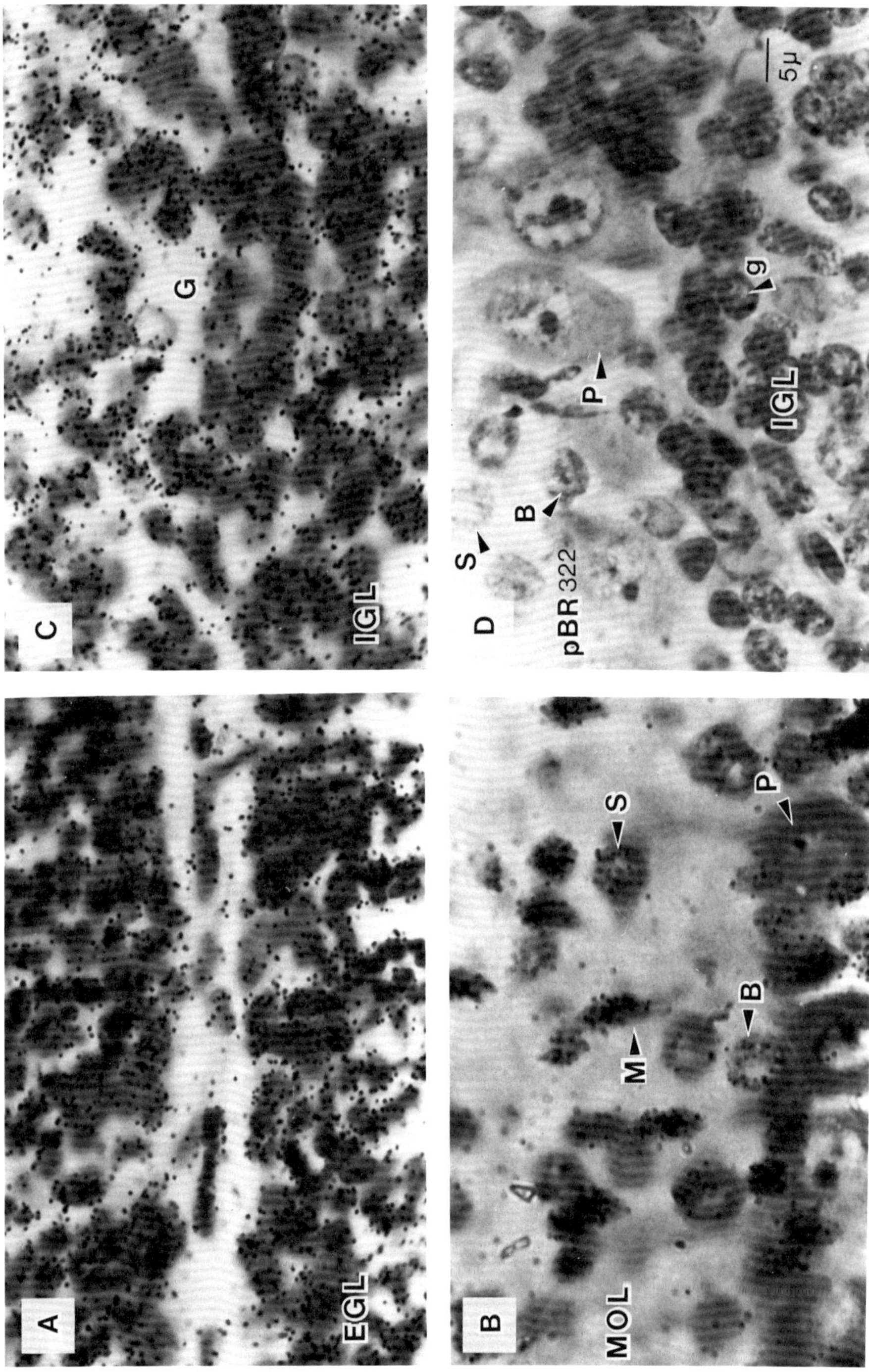

Fig. 2.

TABLE I. In Situ Hybridization of Beta-Tubulin cDNA and [^{3}H]-Poly(U) to 14-Day-Cerebellar Neurons

	Grains per cell	Relative levels of beta-tubulin mRNA
Purkinje cells		
beta-tubulin	14.7 ± 0.3	0.48
poly(U)	30.1 ± 3.4	
Stellate cells		
beta-tubulin	24.7 ± 1.8	0.35
poly(U)	70.6 ± 2.6	
EGL granule cells		
beta-tubulin	53.0 ± 2.0	2.30
poly(U)	23.0 ± 2.4	
IGL granule cells		
beta-tubulin	33.0 ± 2.4	1.36
poly(U)	24.0 ± 3.0	

The distribution of relative levels of tubulin mRNAs and total poly(A)$^+$ mRNAs over Purkinje and stellate cells and the granule cells in the EGL and IGL. Grains were counted over cells such as those shown in Figures 1 and 2. The 5 μm cerebellar sections, examined at 1000 diameters magnification, yielded "grains/cell" (mean ± SD of at least 100 cells of each cell type from three different experiments). Grains were counted by focusing at every level of the emulsion coated slide. The relative level of tubulin mRNA in each cell type is the ratio of grains/cell type in sections hybridized to the ^{3}H-beta tubulin probe to the grains/cell type in sections hybridized to ^{3}H poly(U).

cytoplasmic volume of Purkinje cells is several hundred-fold greater than that of granule cells [7].

In contrast to the cellular contents of total poly(A)$^+$ mRNA, tubulin mRNA levels were by far the highest in granule cells of the EGL (53 grains per cell). Granule cells in the IGL had an average of 33 grains per cell and stellate and Purkinje cells averaged 24.7 and 14.7 grains per cell, respectively. When the ratios of grains hybridizing to poly(A)$^+$ mRNAs and tubulin mRNAs were calculated, it was clear that the relative amount of beta tubulin mRNA was twofold greater in granule cells of the EGL compared to those in the IGL. Moreover, the relative beta tubulin mRNA content of Purkinje cells was 50% less than that of granule cells of the IGL and stellate cells contained only 25% as much.

Fig. 2. Autoradiographs of in situ hybridization of ^{3}H-beta tubulin probe and ^{3}H-native pBR322. ^{3}H beta-tubulin probe (6×10^5 cpm/μg; 60,000 cpm/section) (A–C) and native pBR322 (3×10^6 cpm/μg; 60,000 cpm/section) (D) were hybridized to cerebellar sections from 14-day-old rats. A) Grain density over granule cells in the EGL. B) Grain density in Purkinje cells (P), migrating cells (M) as well as in the stellate (S) and basket (B) cells in the molecular layer. C) Grain density in granule cells of the IGL. D) Grain density in Purkinje cells, stellate and basket cells, as well as in granule cells of the IGL. The low density of grains over cells in D shows the lack of hybridization of radiolabeled native pBR322 to cerebellar cells. Magnification, ×500.

DISCUSSION

The reproducibility of grain counts in individual cell types in these experiments clearly shows that in situ hybridization can be used quantitatively. Others have documented that, even under subsaturating conditions, grain intensity is proportional to RNA content [13,18–22]. However, it is important to note that one cannot exclude the possibility that some cells are affected differentially by tissue processing so that loss of RNA might vary according to cell type. Furthermore, the accessibility of RNAs to the probe might be different in one cell type compared to another. If either or both of these possibilities is true, comparison of the levels of poly(A)$^+$ mRNAs or tubulin mRNAs in different cell types would not be valid. However, in the likely event that poly(A)$^+$ and tubulin mRNAs are equally accessible for in situ hybridization in a single cell type, the amount of tubulin mRNA relative to the amount of poly(A)$^+$ mRNA can be compared in different cell types.

It has been assumed from morphological studies that both stellate and granule cells, with their proportionally large nuclei and organelle-poor cytoplasm, are relatively inactive compared to the Purkinje cells, with their extensive, polysome-rich cytoplasm [7]. However, if total grain counts in the poly(U) sections are an accurate reflection of intracellular poly(A)$^+$ mRNA levels in vivo, then, despite their small size (5–10 μm) [7], stellate cells contain more poly(A)$^+$ mRNA than any of the other neuronal types in 14-day cerebellum. The large Purkinje cells (21 μm) [7] contain much less poly(A)$^+$ mRNA, indeed, approximately the same amount as the small granule cells with their scant cytoplasm [7]. When the volume of cytoplasm per cell is considered, the present results therefore would suggest that the poly(A)$^+$ mRNA concentration in Purkinje cells is many fold less than in either of the other cell types. Preliminary data indicate that the paucity of mRNA in the prenatally formed Purkinje cells reflects their relative maturity at postnatal day 14, whereas the EGL and IGL granule cells have equivalent poly(A)$^+$ mRNA levels at different stages of maturity (unpublished observation). However, variations in the efficiency of hybridization to different cell types would preclude estimation of intracellular mRNA levels. Different fixation and hybridization paradigms are now being compared to determine the optimal conditions for comparison of absolute levels of intracellular mRNAs in the different cell types. A comparison of cellular mRNA levels at different developmental stages then will be necessary to determine whether mRNA levels decrease in individual cell types during development.

The distribution of beta tubulin mRNAs in the cerebellum is different from that of poly(A)+ mRNA, being highest in EGL cells. When the beta tubulin mRNA levels relative to the total poly(A)$^+$ mRNAs levels are compared (thus eliminating possible differences in the efficiency of in situ hybridization and/or differential mRNA leakage from the various cell types, as discussed above), it was found that granule cells in the EGL contain twofold more beta tubulin mRNA than those in the IGL. This could be a partial explanation for the developmental decrease in total beta tubulin mRNA levels that have been observed in both rat [1,9,10] and human [8]. The functional significance of this decrease may be related to the different metabolic requirements of granule cells during mitosis and axon formation in the EGL and during their final maturation in the IGL. In other systems, it has been shown that axon regrowth is accompanied by an increase in tubulin mRNA levels [29], indicating that normal axon growth, particularly in parallel fibers that contain microtubules and no neurofilaments [7], may require similarly increased synthesis of tubulin. Relatively high levels of tubulin mRNAs in the premigratory cell populations of the EGL would thus be expected.

Comparison of the relative levels of tubulin mRNAs in the postmigratory, differentiated IGL granule cells, stellate and Purkinje cells shows three- to fourfold greater levels in the granule cells. This suggests that elaboration of dendrites in the 14-day stellate and Purkinje cells requires less tubulin synthesis than does maintenance of the long parallel fibers by the small cell bodies of the granule cells.

The beta tubulin recombinant DNA probe used in this study is a full-length copy of a chick beta tubulin mRNA [23,30] and thus will hybridize with the coding regions of the mRNAs for the other beta tubulin isotypes [31,32]. However, under our stringent experimental conditions, it is unlikely that the probe hybridizes to any non-beta-tubulin mRNAs. Several lines of evidence indicate that a number of beta tubulin mRNAs are expressed in vertebrate cells [1–8,10,31,32], but only one, the beta 1 subunit, is decreased dramatically during brain development [1,8]. Subcloning of the 3′ ends of the rat beta tubulin mRNAs from our rat cerebellar cDNA library [Morrison and Sparkman, manuscript in preparation] and hybridization of these probes to cerebellar slices should allow us to determine whether the different tubulin isotypes are localized in specific cell types or whether the beta 1 tubulin mRNA is synthesized preferentially in the dividing and premigratory cells. Such an analysis would indicate a specific function for the beta 1 subunit.

ACKNOWLEDGMENTS

This work was supported in part by NIH grants AI 14663 and HD 14886. We wish to thank Chris Claeson for typing the manuscript.

REFERENCES

1. Morrison MR, Pardue S, Griffin WST: J Biol Chem 256:3550, 1981.
2. Dustin P: In "Microtubules." New York: Springer-Verlag, 1978.
3. Nunez J, Francon J, Lennon AM, Fellous A, Mareck A: In DeBrabander M, DeMey J (eds): "Microtubules and Microtubule Inhibitors." Amsterdam: Elsevier/North Holland Biomedical Press, 1975, p 213.
4. Thao NB, Wooten GF, Axelrod J, Kopin IJ: Proc Natl Acad Sci USA 69:520, 1972.
5. Olmstead JB, Borisy GG: Ann Rev Biochem 42:507, 1973.
6. Addison WH: J Comp Neurol 21:459, 1911.
7. Palay S, Chan-Palay V: "Cerebellar Cortex." New York: Springer-Verlag, 1974.
8. Morrison MR, Pardue S, Griffin WST: J Neurogenetics 1:105, 1983.
9. Gozes I, De Baetselier A, Littauer UZ: Eur J Biochem 103:13, 1980.
10. Bond JF, and Farmer SR: Mol Cell Biol 3:1333, 1984.
11. Roberts JL, Chen C-LC, Eberwine JH, Evinger MJQ, Gee C, Herbert E, Schachter BS: Recent Prog Horm Res 38:227, 1982.
12. MacDonald RJ, Swift GH, Quinto C, Swain W, Pictet RL, Nikovits W, Rutter WJ: Biochemistry 21:1453, 1982.
13. Hafen E, Levine M, Garber RL, Gehring WJ: EMBO J 2:617, 1983.
14. Harrison PR, Conkie D, Affara N, Paul J: J Cell Biol 63:402, 1974.
15. Jakowlew SB, Khandekar P, Datta K, Arnold HH, Narula SK, Siddiqui MAQ: J Mol Biol 156:673, 1982.
16. Singer RH, Ward DC: Proc Natl Acad Sci USA 79:7331, 1982.
17. Szabo P, Elder R, Steffensen DM, Uhlenbeck OC: J Mol Biol 115:539, 1977.

18. Brahic M, Haase AT: Proc Natl Acad Sci USA 75:6125, 1978.
19. Angerer LM, Angerer RC: Nucleic Acids Res 12:1819, 1981.
20. Cox KH, DeLeon DV, Angerer LM, Angerer RC: Dev Biol 101: 485, 1984.
21. Lynn DA, Angerer LM, Bruskin AM, Klein WH, Angerer RC: Proc Natl Acad Sci USA 80:2650, 1983.
22. Capco DG, Jeffery WR: Dev Biol 67:137, 1978.
23. Cleveland DW, Lopata MA, MacDonald RJ, Cowan NJ, Rutter WJ, Kirschner MW: Cell 20:95, 1980.
24. Rigby PW, Dieckman J, Rhodes C, Berg P: J Mol Biol 113:237, 1977.
25. Griffin WST, Alejos M, Morrison MR: Brain Res Bull 10:597, 1983.
26. Jeffery WR: J Cell Biol 95:1, 1982.
27. Morrison MR, Brodeur R, Pardue S, Baskin F, Hall CL, Rosenberg RN: J Biol Chem 254:7675, 1979.
28. Venezky DL, Angerer LM, Angerer RC: Cell 24:385, 1981.
29. Ginzburg I, Scherson T, Rybak S, Kimh Y, Neuman D, Schwartz M, Littauer, UZ: CSHQB XLVII: 783, 1983.
30. Valenzuela P, Quiroga M, Zaldivar J, Rutter WJ, Kirschner MW, Cleveland DW: Nature 289:650, 1981.
31. Hall JL, Dudley L, Dooner PR, Lewis SA, Cowan NJ: Mol Cell Biol 3:854, 1983.
32. Lopata MA, Havercroft TC, Chow LT, Cleveland DW: Cell 32:713, 1983.

ADDENDUM

We have recently shown by Northern analysis of cerebellar mRNAs that the chick β tubulin probe encodes the same mRNA as the rat RBT.2 probe of Bond *et al* [Mol Cell Biol 4:1313, 1984] and shows little or no cross hybridization with mRNAs to the more abundant brain β tubulin mRNA which is also expressed in other tissues. Therefore, our in situ hybridization is to the brain-specific β tubulin mRNA, the levels of which decrease dramatically during development of both cerebellum and cortex [1,10].

Journal of Cellular Biochemistry 27:415–422 (1985)
Neurobiology: Molecular Biological Approaches to Understanding Neuronal Function and Development 65–72

In Situ Hybridization of Putative Somatostatin mRNA Within Hypothalamus of the Rat Using Synthetic Oligonucleotide Probes

Rene Arentzen, Frank Baldino, Jr., Leonard G. Davis, Gerald A. Higgins, Yuan Lin, Robert W. Manning, and Betty Wolfson

Central Research & Development Department, E. I. du Pont de Nemours and Company, Wilmington, Delaware 19898

The distribution of mRNA with high sequence homology to somatostatin mRNA within the periventricular hypothalamus of rat was assessed using in situ hybridization techniques with synthetic oligodeoxyribonucleotide probes, complementary to the 3′ coding region of rat somatostatin mRNA. The probes (22- and 24-mers) were 5′-end labeled using T4 polynucleotide kinase and γ-^{32}P-ATP. They were used either individually or after ligation with T4 DNA ligase to form a 46-mer. Serial tissue sections (< 10 μm) were taken from the level of the preoptic/anterior hypothalamus through the paraventricular hypothalamus. In situ hybridizations were conducted at room temperature in hybridization buffer. Neurons immunoreactive with antiserum raised against somatostatin were identified in alternate sections using standard immunocytochemical procedures. The anatomical location of the hybridization signal was determined by autoradiography. Our results show that the peri- and paraventricular hypothalamus is rich in transcripts putatively coding for somatostatin and that these transcripts are co-distributed with neurons immunoreactive with antisomatostatin immunoglobulin.

Key words: in situ hybridization, somatostatin, mRNA, immunocytochemistry, hypothalamus

Somatostatin is a tetradecapeptide, originally isolated from bovine hypothalamus as a modulator of growth hormone release [1]. Since then, neurons recognized by antisomatostatin immunoglobulin have been found not only within neurosecretory cells of the periventricular hypothalamus [2,3] but also in extrahypothalamic regions of the central nervous system [4,5]. These findings suggest that the function of somatostatin may not be limited to the neuroendocrine system.

Immunocytochemistry, however, is not a reliable means of identifying sites of somatostatin synthesis. A better approach is to identify somatostatin mRNA directly in brain tissue sections by in situ hybridization with labeled DNA probes [6,7,8]. The

Authors are arranged in alphabetical order.

Received April 12, 1984; revised and accepted October 24, 1984.

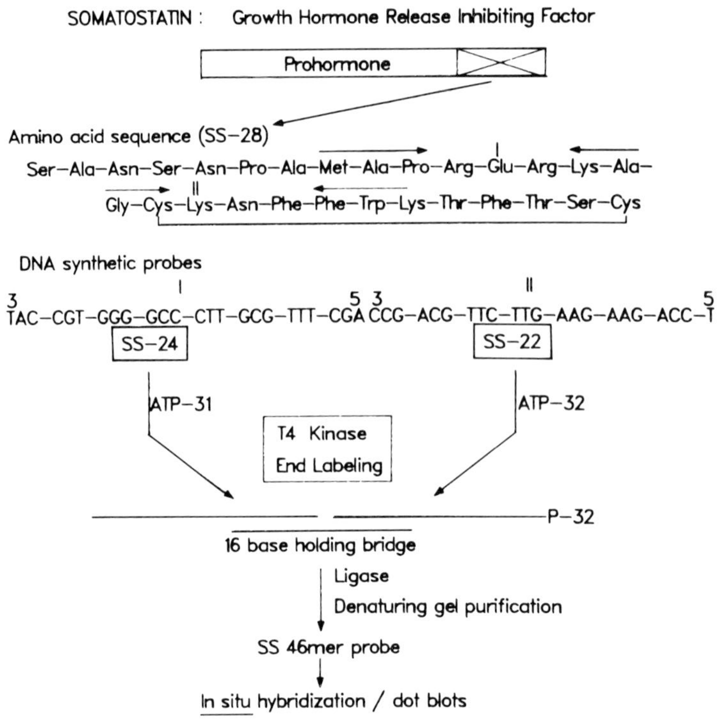

Fig. 1. Schematic illustration of the probe sequence and ligation reaction used to construct the 46-base oligonucleotide probe (SS = somatostatin).

recent determination of a cDNA sequence of preprosomatostatin from rat medullary thyroid carcinoma [9] has made it possible to synthesize specific oligonucleotide probes for use in such experiments.

In this study, the distribution of putative somatostatin mRNA within the periventricular portion of the preoptic area and paraventricular nucleus of the rat was examined using synthetic DNA probes complementary to the 3′ coding region of rat somatostatin mRNA (Fig. 1). The pattern of hybridization within these regions was compared to the distribution of neurons immunoreactive with antiserum which recognizes the C-terminal portion of somatostatin. Throughout this report, the use of the terms somatostatin mRNA and somatostatin immunoreactivity implies their putative nature.

MATERIALS AND METHODS

Preparation of Oligonucleotide Probes

Oligodeoxyribonucleotides (22- and 24-mers) were synthesized on a Model 380A DNA synthesizer (Applied Biosystems, Inc.), and purified by reverse phase HPLC on a μBondapak C_{18} column (Waters Associates) [10]. Oligonucleotides were 5'-end labeled using T4 polynucleotide kinase and γ-^{32}P-ATP [11]. The ^{32}P-labeled probes were either used individually, or after ligation, with the aid of a 16-mer holding-fragment and T4 DNA ligase to form a 46-mer, as shown in Figure 1. All probes had final activities of 1–5 $\times$ 10^8 cpm/μg.

In Situ Hybridization Histochemistry

Hybridization procedures were adopted from those published previously [6–8, 12–15]. Adult Sprague-Dawley rats were perfused transcardially with phosphate-buffered saline (PBS), followed by phosphate-buffered 10% formalin. The brains then were removed and placed in PBS containing 30% sucrose overnight at 4°C. Transverse 5–10 μm sections were cut from the hypothalamus on a cryostat and mounted on slides that had been dipped briefly in sterile diethylpyrocarbonate (50 μl/250 ml H_2O). The slide-mounted sections then were dipped briefly in the diethylpyrocarbonate solution, rinsed twice with sterile, distilled, deionized H_2O and incubated for 15 min at 37°C with proteinase K (1 μg/ml in 20 mM Tris-HCl, 2 mM $CaCl_2$, pH = 7.2). In some experiments, alternate sections were incubated at 37°C for 30 min with ribonuclease I (10 μg/ml in 50 mM Tris-HCl, 10 mM NaCl, and 10 mM EDTA, pH = 8.0). The slides were delipidated with absolute ethanol, rehydrated, and then rinsed in sterile, distilled, deionized H_2O. The washed slides then were placed in a hybridization buffer consisting of 0.3 M NaCl, 0.03 M sodium citrate, (pH = 7.0), 0.2% Ficoll, 0.2% bovine serum albumin, 0.2% polyvinylpyrrolidone, 50% deionized formamide, 0.1% SDS, 0.1% denatured salmon sperm DNA and 0.1% denatured yeast RNA. Twenty μl of ^{32}P-labeled probe (1,000–50,000 cpm in hybridization buffer) was applied to each section for an overnight incubation at room temperature (23°C). The slides were rinsed repeatedly in 0.3 M NaCl, 0.03 M sodium citrate (pH = 7.0), air dried, and exposed to X-ray film (Kodak XAR-5 or SB) for three to seven days. After the X-ray films were processed, the tissue sections were stained with cresyl violet to determine the anatomical location of the hybridization signal.

Immunocytochemical Localization of Somatostatin Immunoreactivity

Alternate sections were mounted on slides and processed for immunocytochemistry with antisomatostatin (Immunonuclear) using the avidin-biotin method [16,17] and diaminobenzidine as a chromogen.

RESULTS

The hybridization signal obtained with the short probes (22- and 24-mers) was distinctly observable along the wall of the third ventricle in the preoptic area (Fig. 2A). More caudally, hybridization also was observed in the paraventricular nucleus. The distribution of somatostatin-immunoreactive neurons within the periventricular region (Fig. 2B) corresponds to the location of the hybridization signal.

Utilization of the 46-base probe permitted reproducible localization of somatostatin mRNA along the wall of the third ventricle from the level of the optic chiasm to that of the ventromedial nucleus. A substantial amount of hybridization was observed within the paraventricular nucleus (Fig. 3). The pattern of hybridization seen with the 46-mer (Fig. 4A) in the periventricular region was comparable to the distribution of somatostatin-immunoreactive perikarya in alternate sections (Fig. 4B). Hybridization with this probe was also observed in the neocortex, pyriform cortex, and the amygdaloid complex, regions previously shown to contain somatostatin-immunoreactive neurons [17]. On the other hand, hybridization was not observed in the cerebellar cortex and regions of the prosencephalon, where somatostatin immunoreactivity has not been identified. In addition, hybridization did not occur in tissue pretreated with ribonuclease I.

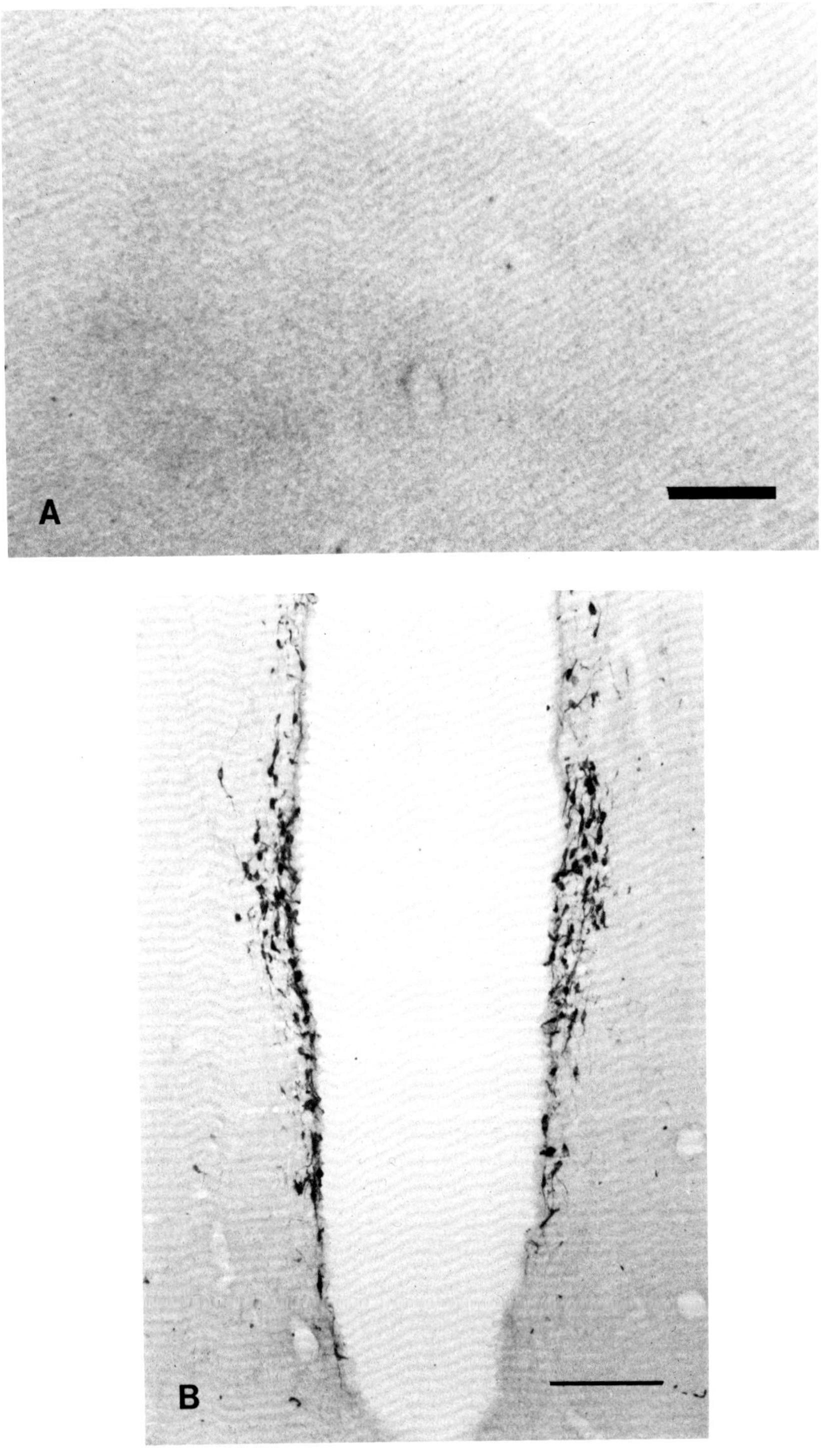

Fig. 2.

DISCUSSION

In this study, we used ^{32}P-labeled synthetic DNA probes to localize somatostatin mRNA within the hypothalamus of the rat. Both the short (22- and 24-mers) probes and the longer 46-mer gave a hybridization signal with the greatest intensity in the periventricular hypothalamus, an area that contains a high density of somatostatin-immunoreactive neurons. The absence of hybridization signal in tissue sections pretreated with ribonuclease I confirmed that the hybridization was specific for RNA. The inability to detect hybridization in tissue regions where somatostatin immunoreactivity was not observed suggests that the observed signal was specific for somatostatin mRNA.

The 46-nucleotide probe gave the best resolution of the three probes. It consistently hybridized to mRNA in the periventricular preoptic area as well as in the paraventricular nucleus. In contrast, the shorter probes produced less robust results, with hybridization often limited to the preoptic area. The difference in the signal-to-

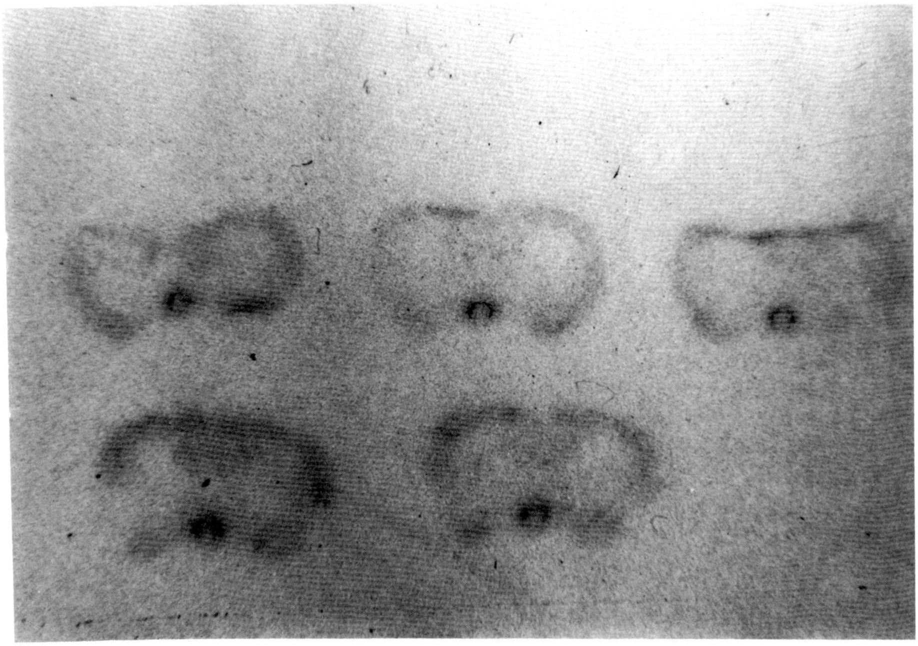

Fig. 3. In situ hybridization with a ^{32}P-labeled 46-base synthetic probe for somatostatin mRNA. Autoradiograph of serial 5 μm tissue sections (50,000 cpm/section), including the paraventricular nucleus. Note the hybridization signal in the periventricular portion of this nucleus.

Fig. 2. In situ hybridization with a 24-base synthetic probe for somatostatin mRNA. A) Autoradiograph of a tissue section (5 μm) incubated with ^{32}P-labeled 24-mer (2000 cpm/section). Note the hybridization signal along the wall of the third ventricle and in the neocortex, particularly the pyriform lobe. (Similar results have been obtained with the 22-mer) Bar = 2mm. B) Somatostatin-immunoreactive cells, surrounding the third ventricle of an adult rat, detected in an alternate 5 μm section by peroxidase-antiperoxidase immunocytochemistry. Bar = 200μm.

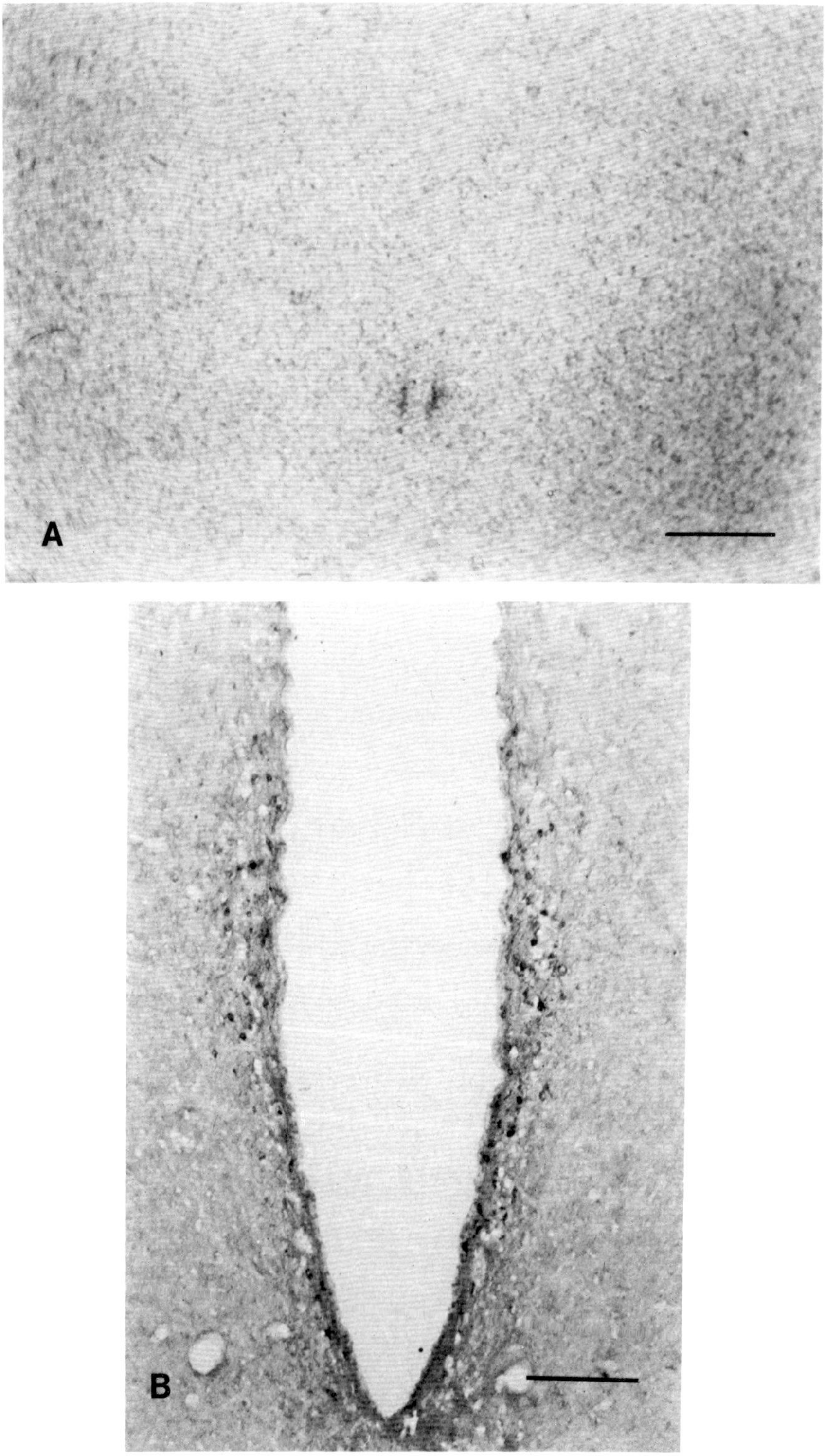

Fig. 4.

noise ratio observed between the short and long probes might reflect differences in mRNA levels between these regions.

Although the coexistence of somatostatin mRNA and somatostatin immunoreactivity within the hypothalamus, or even within individual hypothalamic neurons, implies a transcript-product relationship, it is not proof of such a relationship. Cells and neurons may actively take up peptides or contain untranslated mRNA. Other authors [6,7] have made similar implications concerning the sites of biosynthesis of proopiomelanocortin-related products within the pituitary.

Although these data indicate discrete anatomical sites for somatostatin biosynthesis, they provide no evidence as to the final active peptide synthesized within these regions, as somatostatin-14 and somatostatin-28 are derived from the same prohormone [18]. The presence of hybridization signal within the peri- and paraventricular hypothalamus nevertheless suggests that the transcripts necessary to code for these peptides is present within these hypothalamic regions.

In conclusion, the use of a rapid autoradiographic method to identify somatostatin mRNA in discrete brain regions has been demonstrated in this study. Similar achievements in the pituitary using ^{32}P-labeled cDNA probes have been described previously [8]. One important consequence of these techniques may be the ability to study the regulation of neuropeptide biosynthesis in discrete brain regions at the level of mRNA and, perhaps ultimately, the gene.

ACKNOWLEDGMENTS

The authors wish to acknowledge the technical assistance of M.T. Moke, T. O'Kane, and C. A. Phillips. The authors are grateful to Dr. S. J. Watson for advice and helpful discussions.

REFERENCES

1. Brazeau P, Vale W, Burgus R, Ling N, Butcher M, Rivier J, Guillemin, R: Science 179:77–79, 1973.
2. Hökfelt T, Efendic S, Johansson O, Luft R, Arimura A: Brain Res 80:165–169, 1974.
3. Sawchenko PE, Swanson LW: J Comp Neurol 205:260–272, 1983.
4. Brownstein M, Arimura A, Sato H, Schaly AV, Kizer JS: Endocrinology 96:1456–1461, 1975.
5. Vale W, Brazeau P, Rivier C, Brown M, Ross B, Rivier J, Burgus R, Lign N, Guillemin R: Recent Prog Horm Res 31:365–397, 1975.
6. Gee CE, Roberts JL: DNA 2:157–163, 1983.
7. Gee CE, Chen C-LC, Roberts JL, Thompson R, Watson SJ: Nature 306:374–376, 1983.
8. Hudson P, Penschow J, Shine J, Ryan G, Niall H, Coglan J: Endocrinology 108:353–356, 1981.
9. Goodman RH, Aron DC, Roos BA: J Biol Chem 258:5570–5573, 1983.
10. Fritz HJ, Belagaje R, Brown EL, Fritz RH, Jones RA, Lees RG, Khorana HG: Biochemistry 17:1257–1267, 1978.

Fig. 4. In situ hybridization with a 46-base oligonucleotide probe for somatostatin mRNA. A) Autoradiograph of a tissue section (7 μm) incubated with ^{32}P-labeled 46-mer (2000 cpm/section). This section includes the paraventricular nucleus. The hybridization signal is limited to the region surrounding the third ventricle. Bar = 2mm. B) Somatostatin-immunoreactive neurons along the wall of the third ventricle detected by the avidin-biotin method in an alternate 7 μm section. Bar = 200μm.

11. Maniatis T, Fritsch EF, Sambrook J (eds): 'Molecular Cloning: A Laboratory Manual.' Cold Spring Harbor: Cold Spring Harbor Laboratory, 1982, p 125.
12. Gall JG, Pardue ML: Proc Natl Acad Sci USA 63:378–383, 1969.
13. John H, Birnstiel ML, Jones KW: Nature (Lond) 223:582–587, 1969.
14. Szabo P, Ward DC: Trends in Biochem Sci 7:425–427, 1982.
15. Brahic M, Haase AT: Proc Natl Acad Sci USA 75:6125–6129, 1978.
16. Hsu SM, Raine L, Fanger H: J Histochem Cytochem 29:577–580, 1981.
17. Higgins GA, Schwaber JS: Peptides 4:657–662, 1983.
18. Itallie CM, Fernstrom JD: Endocrinology 113:1810–1217, 1983.

Journal of Cellular Biochemistry 27:423–441 (1985)
Neurobiology: Molecular Biological Approaches to Understanding Neuronal Function and Development 73–91

Expression and Assembly of the Erythroid Membrane-Skeletal Proteins Ankyrin (Goblin) and Spectrin in the Morphogenesis of Chicken Neurons

Elias Lazarides and W. James Nelson

Division of Biology, California Institute of Technology, Pasadena, California 91125

The membrane-skeleton of adult chicken neurons in the cerebellum and optic system is composed of polypeptides structurally and functionally related to the erythroid proteins spectrin and ankyrin, respectively. Neuronal spectrin comprises two distinct complexes that share a common α subunit (M_r 240,000) but which have structurally distinct polymorphic subunits ($\beta'\beta$ spectrin; M_r 220/225,000; γ spectrin, M_r 235,000); the brain-specific form ($\alpha\gamma$ spectrin or fodrin) and an erythrocyte-specific form ($\alpha\beta'\beta$ spectrin). Two structurally related isoforms of ankyrin have also been identified and are termed α (M_r 260,000) and β (M_r 237,000) ankyrin. Immunofluorescence demonstrates that the variants of spectrin and ankyrin, respectively, have different distributions within neurons. On the one hand, $\alpha\gamma$ spectrin and β ankyrin are present throughout the neuron, in the perikaryon, dendrites, and axon, whereas $\alpha\beta'$ spectrin and α ankyrin are localized exclusively in the perikaryon and dendrites where they are actively segregated from $\alpha\gamma$ spectrin and other components of axonal transport. This asymmetric distribution of spectrin and ankyrin isoforms is established in distinct stages during neuronal morphogenesis. Early in cerebellar and retinal development, $\alpha\gamma$ spectrin is expressed in mitotic cells. Subsequently β ankyrin and $\alpha\gamma$ spectrin are co-expressed in postmitotic cells and gradually accumulate on the plasma membrane in a uniform pattern throughout the neuron during the phase of cell growth. At the onset of synaptogenesis and the cessation of cell growth, their levels of synthesis decline sharply while the assembled proteins remained as stable membrane components. Concomitantly, there is a dramatic induction in the accumulation of α ankyrin and $\alpha\beta'$ spectrin, whose assembly is limited to the plasma membrane of the perikarya and dendrites. These results demonstrate that two successive, developmentally regulated programs of ankyrin and spectrin expression and patterning on the plasma membrane are involved in the assembly of the spectrin-based asymmetry in the neuronal membrane-skeleton, and that their asymmetric distribution is actively maintained throughout the life of the neuron.

Key words: ankyrin, spectrin, membrane assembly, neuronal differentiation

W. James Nelson's present address is Institute for Cancer Research, Fox Chase Cancer Center, Philadelphia, PA 19111.

Received September 25, 1984; accepted December 4, 1984.

Morphological and electrophysiological studies on neurons in vivo and in vitro have demonstrated a striking structural and physiological asymmetry in these cells. At one end of the neuron is the perikaryon and associated dendrites, which receive electrical and chemical signals that impinge onto the postsynaptic terminals located on their plasma membranes; the perikaryon also contains the nucleus and all the protein-synthesizing machinery of the cell. These signals are relayed down the axon, which extends away from the perikaryon, by the unidirectional propagation of action potentials and the axonal transport of proteins and organelles from the perikaryon to the distal end of the axon, where the presynaptic terminal is located. Hence, the asymmetry of the neuron is defined by the morphological and physiological differences that exist between the perikaryon, dendrites, and axon, respectively. This asymmetry is established during neuronal morphogenesis in at least two discrete stages. The first stage involves the outgrowth of the axon and dendrites, which occurs during the phase of cell growth following withdrawal of the cell from the proliferative cell cycle. Subsequently, upon contact of the growth cone of one neuron with the perikaryon or dendrites of another (synaptogenesis), cell growth ceases and the phase of terminal differentiation is initiated. The concomitant gradual formation of the pre- and postsynaptic terminals upon synaptogenesis marks the second stage in the establishment of the structural and physiological asymmetry of the neuron.

Although the molecular mechanisms which regulate these morphogenetic events are largely unknown, the striking structural asymmetry of the mature neuron suggests that the assembly of the membrane-skeleton may play an important role in neuronal morphogenesis. Furthermore, since neurons, like many other cell types, contain a subcortical cytoskeletal network of actin filaments, a mechanism for restricting the distribution of proteins to different domains on the plasma membrane in the mature neuron may also involve elements of the cytoskeleton and, in particular, the membrane-associated forms of actin and actin-binding proteins that mediate the attachment of actin to the plasma membrane.

To gain an insight into the molecular basis of neuronal morphogenesis and the role of the membrane-skeleton in the maintenance of the structural and physiological asymmetry of the neuron, we have recently analyzed biochemically and immunologically the molecular composition of the membrane-skeleton of neurons in the developing and adult chicken cerebellum and optic system. As a basis for these studies, we have sought to identify specific polypeptides that are structurally and functionally related to the well-defined components of the erythrocyte membrane-skeleton, in which the molecular interactions between integral membrane proteins and structural proteins in the subcortical cytoplasm have been characterized extensively.

In the erythrocyte, the membrane-skeleton consists predominantly of actin and the actin-binding protein spectrin, the latter accounting for 75% of the protein mass [50]. Spectrin is composed of two nonidentical polypeptides, termed α spectrin (M_r 240,000) and β spectrin (M_r 220,000) that self-associate to form an $(\alpha\beta)_2$ tetramer [for references, see 8,16]. This spectrin tetramer binds to and crosslinks filamentous (F) actin [17] in the subcortical cytoplasm and, through the β subunit, is attached to another polypeptide, termed ankyrin, which is itself directly bound to the transmembrane-orientated anion transporter [9–11,75]. In avian erythrocytes, the structural and functional analogue of ankyrin is a protein termed goblin [53,56], which has been characterized previously as a phosphoprotein whose phosphorylation is hormone-sensitive and is correlated quantitatively with an increase in Na^+ -K^+ -cotransport

[2,3,7]. Additionally, spectrin tetramers may interact with the plasma membrane through a polypeptide termed protein 4.1 [for references see 8,16,46]. Protein 4.1 interacts simultaneously with spectrin and actin and greatly increases the affinity of the latter two polypeptides for each other [for references see 8,16] and further mediates their attachment to the plasma membrane by interacting with the membrane-bound protein glycophorin [4]. These specific protein interactions form a highly crosslinked protein matrix which provides tensile strength to the plasma membrane and restricts the lateral mobility of certain membrane proteins in the plane of the membrane.

The polypeptides identified in our studies on the neuronal membrane-skeleton are the analogues in chicken neurons of chicken erythrocyte spectrin and ankyrin (goblin). Using a combination of biochemical and immunological techniques, we have demonstrated the expression of discrete isoforms of spectrin and ankyrin in cerebellar and retinal neurons, and the existence of a spectrin-based asymmetry in the molecular composition of the membrane-skeleton of these two neuronal systems. Furthermore, by analyzing the synthesis and accumulation of ankyrin and spectrin during development we have demonstrated that the assembly of different isoforms occurs in distinct steps during neuronal morphogenesis. Together, the results obtained provide for the first time experimental evidence to support the hypothesis that the membrane-skeleton plays an important role in neuronal morphogenesis and in the maintenance of the morphological and physiological asymmetry of the mature neuron.

IDENTIFICATION OF COMPONENTS OF THE MEMBRANE-SKELETON IN ADULT CHICKEN NEURONS

Two Forms of Spectrin are Expressed in Neurons of Adult Chicken Cerebellum and Optic System

Immunoautoradiography of SDS-solubilized extracts of adult chicken cerebellum with a polyclonal antiserum specific for the α subunit of chicken erythrocyte spectrin has revealed the presence of a polypeptide with a apparent molecular weight (M_r 240,000) and isoelectric point similar to that of erythrocyte α spectrin [12,19,29,62,64]. Two-dimensional peptide mapping of chymotryptic iodopeptides derived from chicken erythrocyte α spectrin and from this serologically related polypeptide in chicken cerebellum has demonstrated that these polypeptides are structurally indistinguishable, indicating that they share homologous, if not identical, primary structures [31,58]. However, mammalian erythrocyte α spectrin has a different two-dimensional peptide map to the serologically related polypeptide from mammalian brain [12], indicating that mammalian erythrocyte α spectrin has diverged structurally during evolution from its counterpart in brain and avian erythrocytes.

Under certain experimental conditions, the chicken α spectrin antiserum coimmunoprecipitates from cerebellar extracts a second polypeptide (M_r 235,000) in equimolar amounts with α spectrin [48,55], which is both serologically and, by peptide mapping, structurally distinct from both the α and the β subunits of erythrocyte spectrin [31,55,58]; this polypeptide is termed γ spectrin [59] (see Fig. 1). Several studies have shown that brain α spectrin and γ spectrin copurify with neuronal plasma membranes as a complex whose conformation is similar to that of erythrocyte $\alpha\beta$ spectrin; this polypeptide has been referred to in the past as fodrin or brain spectrin [12,48]. The brain aγ-spectrin complex is a rod-shaped molecule comprising

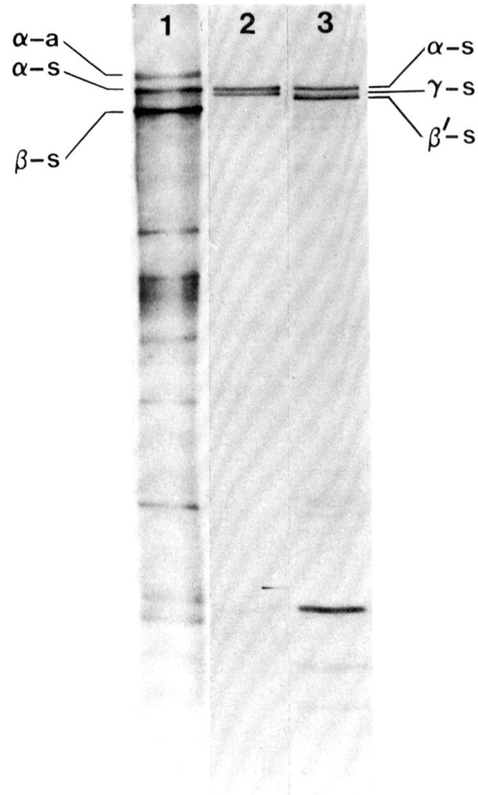

Fig. 1. Identification of two distinct spectrin complexes in neurons of adult chicken retina. A 2-day-old chicken was anesthetized with halothane (1–2%) mixed with nitrous oxide/oxygen (2:1). Approximately 1 mCi [^{35}S]methionine in 10 μl phosphate-buffered saline was pressure-injected into one eyeball using a glass micropipette. The chicken was sacrificed 96 hr after injection. The retina was removed and homogenized in an all-glass Dounce homogenizer in 2 ml of 0.5% Empigen, 10 mM Tris-HCl, pH 7.5, 2 mM EDTA, 1 mM EGTA, 1 mM DTT, 130 mM NaCl, and then stirred at 4°C for 2 hr. Insoluble material was removed by centrifugation at 50,000g for 10 min. The supernatant was divided into two fractions; 2 μl chicken erythrocyte α-spectrin antiserum [64] was added to one fraction and 2 μl chicken erythrocyte β-spectrin antiserum [55] was added to the second. The immunoprecipitation reaction was allowed to proceed overnight at 4°C. The immune complexes from each fraction were isolated with protein A-bearing Staphylococcus aureus, and analyzed by SDS-12.5% polyacrylamide gel electrophoresis and fluorography. The resulting fluorogram is shown here. Lane 1) ^{35}S-methionine-labeled total membrane-skeletal proteins from 10-day chick embryo erythroid cells. Lane 2) α-spectrin immunoprecipitate of ^{35}S-methionine-labeled retina. Lane 3) β-spectrin immunoprecipitate of ^{35}S-methionine-labeled retina. α-s, α spectrin; γ-s, γ spectrin; β′-s, β′ spectrin; β-s, β spectrin; α-a, α ankyrin.

a tetramer formed by the head-on association of two identical dimers, composed of a 1:1 molar ratio of α:γ spectrin [12,30]. Furthermore, the brain spectrin tetramer also binds to and crosslinks F actin [12,30].

Recently, we have demonstrated that neurons of the adult chicken cerebellum and optic system also contain an erythroid form of spectrin [43,44,45]. This second form of spectrin in neurons was identified using a polyclonal antiserum specific for the β subunit of chicken erythrocyte spectrin [55]. Previous studies had shown that this antiserum reacts specifically with chicken erythrocyte β spectrin (M_r 220,000) and a minor component (M_r 230,000) of chicken erythrocyte plasma membranes with which β spectrin shares antigenic determinants but which is antigenically and structurally distinct from α spectrin, γ spectrin, and ankyrin (goblin); this polypeptide is termed β' spectrin [55,58]. Immunoautoradiography of SDS-solubilized extracts of adult chicken cerebellum [43] and retina [45] with the β-spectrin-specific antiserum has revealed the presence of two crossreacting polypeptides. The predominant component has an electrophoretic mobility similar to that of erythrocyte β' spectrin, while the minor component has a slightly faster relative mobility which is similar to that of β spectrin. Two-dimensional mapping of chymotryptic iodopeptides has demonstrated a striking homology between neuronal and erythroid β' spectrin, indicating that they have similar, if not identical, primary structures [45]. The relatively low amounts of the minor component of neurons that crossreacts with the β-spectrin antiserum has, thus far, precluded a similar structural comparison of this polypeptide with its erythroid counterpart.

Immunoprecipitation with β-spectrin antiserum of adult chicken cerebellar and retinal proteins extracted in the presence of the Zwitterionic detergent Empigen demonstrates that β' spectrin is present as a complex with equimolar amounts of α-spectrin, which is coimmunoprecipitated with β' spectrin under these conditions (Fig. 1). Little or no γ spectrin is coimmunoprecipitated with the $\alpha\beta'$-spectrin complex (Fig. 1).

Taken together, these results demonstrate clearly that two forms of spectrin are expressed in adult chicken neurons, a brain-specific form, $\alpha\gamma$ spectrin, and an erythroid form, $\alpha\beta'$ spectrin (Fig. 1). Furthermore, $\alpha\gamma$ and $\alpha\beta'$ spectrin appear to be distinct complexes which share a common α subunit.

Expression of Two Ankyrin (Goblin) Isoforms in the Neurons of Adult Chicken Cerebellum and Retina

In view of the fact that an erythroid form of spectrin is present in chicken neurons, we have sought to determine whether polypeptides serologically and structurally related to erythrocyte ankyrin are also present in neurons [57]. Indeed, a polyclonal antiserum raised against erythrocyte ankyrin (goblin) [56] was shown to react with two polypeptides in SDS-solubilized extracts of adult chicken cerebellum and retina [57]. The larger polypeptide has an apparent molecular weight of 260,000 (termed α ankyrin), which is similar to that of the predominant form of ankyrin (goblin) in chicken erythrocytes. The smaller polypeptide has an apparent molecular weight of 237,000 (termed β ankyrin) similar to that of a minor component of chicken erythrocyte membranes with which α ankyrin shares antigenic determinants, but which is antigenically distinct from α, γ and β spectrin [56,57]. Both α and β ankyrin in neurons are characterized by hormone-dependent phosphorylation and copurify with a plasma membrane fraction from brain [57], similar to their counterparts in erythrocytes [2,7,56,57]. In addition, two-dimensional peptide mapping has revealed extensive homologies between the erythroid and neuronal polypeptides, indicating that they have similar primary structures [57]. Thus, these results demonstrate that the ankyrin polypeptides in chicken cerebellum and retina are both structurally and biochemically related to erythrocyte ankyrin.

ASYMMETRY IN THE MOLECULAR COMPOSITION OF THE NEURONAL MEMBRANE-SKELETON

$\alpha\gamma$ and $\alpha\beta'$ Spectrin Have Different Distributions in the Same Neuron

The distribution of $\alpha\gamma$ and $\alpha\beta'$ spectrin in the neurons of adult chicken cerebellum and optic system has been determined by indirect immunofluorescence microscopy [43,45]. Immunofluorescence on frozen sections of cerebellum with the α-spectrin-specific antiserum has revealed that this antigen is present in all three layers of the cerebellar cortex and in all discernible perikarya, dendrites, and axonal processes that populate the molecular, Purkinje cell, and granular cell layers [43]. Since γ spectrin is present in an equimolar complex with α spectrin (see above), we assume that γ spectrin has an identical distribution to that of α spectrin (see also 48).

In contrast to this uniform distribution of $\alpha\gamma$ spectrin throughout the cerebellar cortex, immunofluorescence with the β-spectrin-specific antiserum has revealed that this antigen has a more restricted distribution. Comparison of alternate sections stained with β-spectrin antiserum and with Nissl stain has revealed that the perikarya of granular cells and Purkinje cells show bright fluorescence in association with the plasmalemma, which, in the case of the Purkinje cells, is also detected in the initial portion of their dendritic trunks that extend into the molecular layer [43]. The rest of the molecular layer does not show any substantial fluorescence above that of the preimmune β-spectrin serum, indicating that $\alpha\beta'$ spectrin is not present in the numerous axonal and dendritic processes that populate this layer of the cerebellum. In addition, processes in the white matter were shown to be uniformly negative with the β-spectrin antiserum, indicating that the axonal processes of mossy and climbing fibers and those of Purkinje cells do not contain $\alpha\beta'$ spectrin but do contain $\alpha(\gamma)$ spectrin (see above). A notable exception in the white matter were the perikarya of neurons of the cerebellar nuclei, which stained strongly with the β-spectrin antiserum. As in the case of the Purkinje cells, β-spectrin-specific fluorescence was associated with the periphery of these cells and was also detected a short distance into the main processes emanating from the perikarya [43].

These results demonstrate that $\alpha\gamma$ spectrin is present throughout the neuron in all discernible types of neurons, whereas the distribution of $\alpha\beta'$ spectrin is restricted to the perikarya of certain cerebellar neurons. That the differential distribution of these two forms of spectrin is a general feature of neurons and is not restricted to those present in the cerebellum was demonstrated by our analysis of the distribution of $\alpha\gamma$ and $\alpha\beta'$ spectrin in the neurons of the adult chicken optic system [45]. As in the cerebellum, α spectrin was detected by immunofluorescence in all discernible layers of the retina including the axons of the optic nerve fibers; the perikarya of ganglion cells; numerous processes in the inner plexiform layer, which are presumed to be ganglion cell dendrites and amacrine cell processes; the perikarya of amacrine, bipolar, and horizontal cells in the inner nuclear layer; cell processes in the outer plexiform layer; and the perikarya of rods and cones in the outer nuclear layer. A notable exception was the absence of any reactive forms in the bacillary layer, which contains the outer segments of the rods and cones [45].

Immunofluorescence with the β-spectrin-specific antiserum on frozen sections of retina revealed that, as in the cerebellum, $(\alpha)\beta'$ spectrin has a restricted distribution. β'-spectrin fluorescence was detected at the plasma membrane of only the perikarya of ganglion cells, the axonal processes of optic nerve fibers within the retina (but not in the optic nerve), and a subset of the processes in the inner plexiform layer close to the ganglion cell layer, which are presumed to be the dendritic processes of ganglion cells that extend into the inner plexiform layer in the avian retina [5,60]. The perikarya and processes of amacrine, bipolar, and horizontal cells were uniformly negative. A notable exception, however, were the perikarya of rods and cones in the outer plexiform layer which were positive with β-spectrin antiserum [45].

Distinct Membrane-Skeleton Domains in Neurons

The finding that there is a differential distribution of $\alpha\gamma$ and $\alpha\beta'$ spectrin in adult chicken cerebellar and retinal neurons demonstrates for the first time a distinctive asymmetry in the molecular composition of the membrane-skeleton of neurons, which appears to exist between the perikaryon and dendrites on the one hand and the axon on the other. More recently, an asymmetric distribution in neurons of several other structural proteins has been noted. These include the microtubule-associated proteins -1 and -2 (MAP-1 and MAP-2), of which MAP-1 is present throughout the neuron, whereas MAP-2 is localized to the perikaryon and dendrites [26,38]; the neurofilament subunits, of which the M_r 68,000 and M_r 170,000 subunits are present throughout the neuron, whereas the M_r 195,000 subunit is localized in the axon [37; see also 78]; and isoforms of tubulin, which show a graded distribution between the perikaryon and the distal end of the axon [33]. These results show that the asymmetric distribution of $\alpha\gamma$ and $\alpha\beta'$ spectrin in neurons is not unique but is a feature of several other structural proteins as well, which together suggest that distinct domains exist in the neuronal membrane-skeleton.

ESTABLISHMENT OF ASYMMETRY IN THE MEMBRANE-SKELETON DURING NEURONAL MORPHOGENESIS

The fact that there is an asymmetry in the molecular composition of the neuronal membrane-skeleton raises the question as to the sequence of developmental events involved in its assembly during neuronal morphogenesis, particularly with respect to the phases of cell growth and terminal differentiation. In the chick embryo, neuronal morphogenesis occurs in two major phases. First, upon withdrawal of the cells from the proliferative cell cycle, there is a dramatic outgrowth of their axons and dendrites; this constitutes the phase of cell growth. Subsequently, on about embryonic day 15 in both retinal and cerebellar neurons, the cells commence synaptogenesis and begin the phase of terminal differentiation; at the same time there is a gradual cessation of cell growth [for references on cerebellum, see 40,54,71; and retina, see 5,18, 39,44,60,61, 63,70,73] (see Fig. 2A).

By analyzing the temporal and spatial expression during neuronal morphogenesis of each spectrin and ankyrin isoform, we sought to determine the steps in the assembly of these proteins during the phases of cell growth and terminal differentiation.

Developmentally Regulated Expression and Patterning on the Neuronal Plasma Membrane of Spectrin and Ankyrin Isoforms During Ontogeny of Chick Cerebellum and Retina

To determine the temporal and spatial accumulation of spectrin and ankyrin isoforms during neuronal morphogenesis, a combination of immunofluorescence and biochemical techniques has been used in parallel on the developing chick cerebellum and retina [44,45,57].

For immunofluorescence, frozen sections of the cerebellar cortex from chicken embryos at embryonic days 17–21 were used. The chick cerebellum at these stages of embryogenesis is ideal for the analysis of neurons at all phases of differentiation. In the same frozen section, cells in the following phases can be detected, from the outside to the inside of the cortex: mitotic cells in the external granular cell layer; postmitotic cells (mostly granule cells) migrating through the molecular layer; and Purkinje and granule cells which have reached their designated position in the cortex (Purkinje cell layer and internal granular cell layer, respectively), and which are undergoing terminal differentiation upon the establishment of synaptic connections with climbing and mossy fibers, respectively [40]. For biochemical analysis of the expression of spectrin and ankyrin isoforms during neuronal morphogenesis, both the accumulation at steady state and the level of synthesis were determined. The trend in accumulation was analyzed by immunoautoradiography of SDS-solubilized extracts of cerebellum and retina and the levels of synthesis by ^{35}S-methionine incorporation in vitro into whole retina followed by immunoprecipitation and fluorography (see Fig. 2, and [57]).

Fig. 2. Developmentally regulated changes in the levels of synthesis and rates of accumulation of ankyrin and spectrin isoforms during neuronal morphogenesis. A) Changes in the average widths of layers in the cerebellar cortex of the chick during embryogenesis; the width of the external granular (EGL) layer is an indication of the proportion of mitotic cells, whereas that of the internal granular (IGL) layer indicates the proportion of terminally differentiated cells; ML, molecular layer: plotted from data in Saetersdal [66] and Addison [1]; see also Jacobson [40]. B) Developmentally regulated decline in the level of $\alpha\gamma$-spectrin synthesis during retinal development. Whole retinas from embryonic day 12–21 chicks were labeled radioactively with ^{35}S-methionine in vitro. $\alpha\gamma$ Spectrin was co-immunoprecipitated with α-spectrin-specific antiserum and analyzed by SDS-12.5% polyacrylamide gel electrophoresis and fluorography. Densitometric analysis of the fluorogram was performed and the resulting values expressed as a percentage of that on embryonic day 12. C and D) Developmentally regulated accumulation of ankyrin and spectrin isoforms during chick cerebellar ontogeny. SDS-solubilized extracts of embryonic cerebellum (days 13–21) were separated on SDS-12.5% polyacrylamide gels, transferred to nitrocellulose, and immunoblotted with ankyrin, α- and β-spectrin antisera, respectively. The immunocomplexes were localized with ^{125}I-labeled protein A and subsequent autoradiography. Densitometric analysis of the autoradiograms was performed and the values normalized to the DNA concentration of the initial extract and then exposed as a percentage of the amount present on embryonic day 13. $\alpha\gamma$-s, $\alpha\gamma$ spectrin (○-○); β-s, β spectrin (●-●); α-a, α ankyrin (▲-▲); β-a, β ankyrin (△-△). Arrow indicates day of hatching (H). Data from [57].

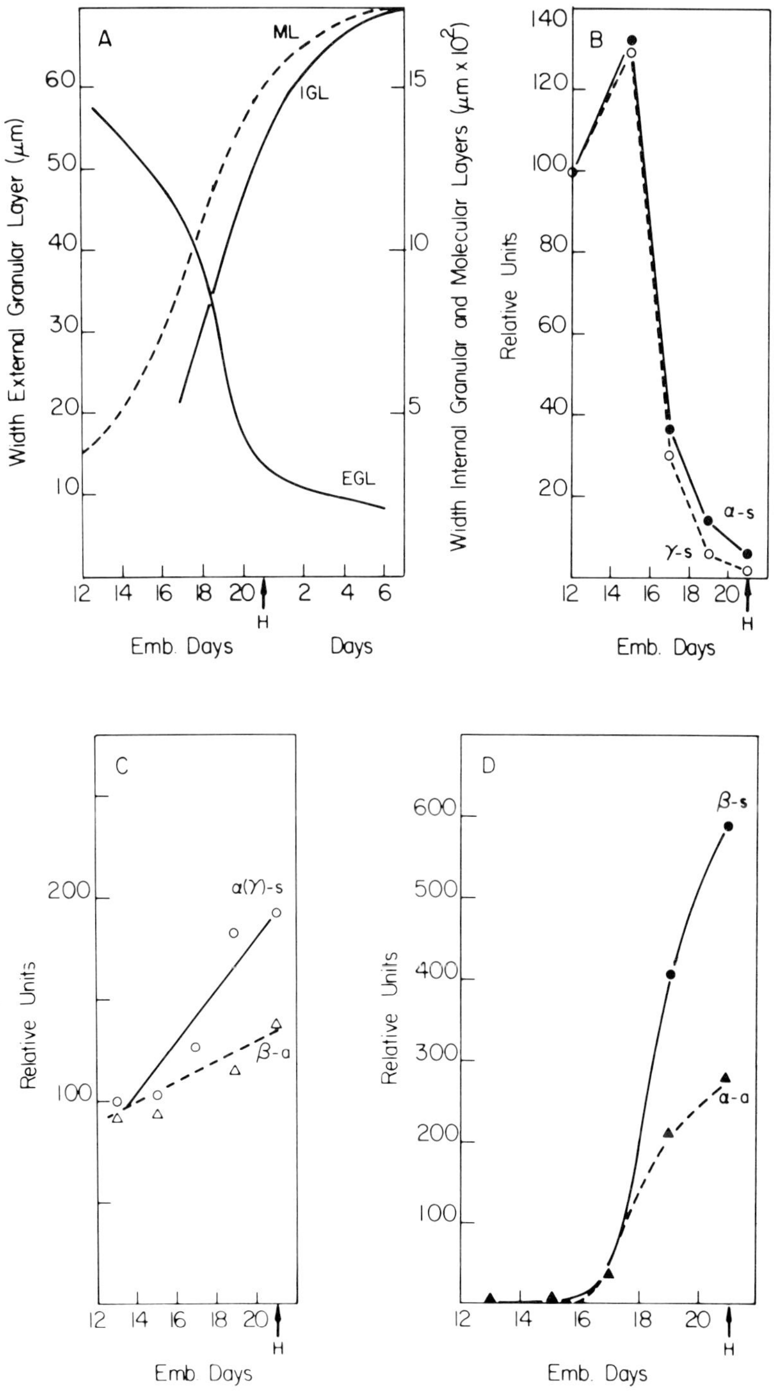
A
ML
IGL
EGL
Width External Granular Layer (μm)
Width Internal Granular and Molecular Layers (μm x 10²)
60
50
40
30
20
10
15
10
5
12 14 16 18 20 2 4 6
H
Emb. Days
Days
B
140
120
100
80
60
40
20
Relative Units
α-s
γ-s
12 14 16 18 20
H
Emb. Days
C
200
100
Relative Units
α(γ)-s
β-a
12 14 16 18 20
H
Emb. Days
D
600
500
400
300
200
100
Relative Units
β-s
α-a
12 14 16 18 20
H
Emb. Days

Fig. 2.

Mitotic cells of embryonic chicken cerebellum and retina express and accumulate at steady state $\alpha\gamma$ spectrin, but not $\alpha\beta'$ spectrin and little or no ankyrin, as shown by indirect immunofluorescence [44,45,57]. However, in the same frozen sections, postmitotic cells exhibit fluorescence with the ankyrin antiserum. This suggests that there is an induction in the expression and/or accumulation of ankyrin upon withdrawal of the cells from the proliferative cell cycle. Immunoautoradiography of a day-13 chick embryo cerebellum, which contains mostly mitotic and postmitotic cells [40], has revealed that the ankyrin isoform expressed at this stage is β ankyrin and that little or no α ankyrin is present [57]. A similar result was obtained from the retina at the same embryonic stage (our unpublished results).

Analysis of the levels of synthesis of $\alpha\gamma$ spectrin in the retina has shown an increase between embryonic days 12 and 15. Subsequently, a sharp decline in the levels was detected such that by the time of hatching the level of $\alpha\gamma$ spectrin synthesis was reduced to approximately 5% of the level observed on embryonic day 12 (Fig. 2B) [see 57]. In contrast, immunoautoradiography of embryonic cerebella and retinas demonstrates a gradual increase in the steady state amount of $\alpha\gamma$ spectrin (Fig. 2C) [see 57], indicating that as the level of synthesis declines there is a concurrent stabilization of newly synthesized protein against proteolytic degradation. During the same period of cerebellar and retinal development, a gradual accumulation of β ankyrin is also detected (Fig. 2C). Although ^{35}S-methionine incorporation into β ankyrin was detected during retinal ontogeny, the amounts were too low to quantify. However, we were able to observe a qualitative decline in the incorporation of ^{35}S-methionine into β ankyrin between embryonic day 13 and day 21; in the adult, little or no β-ankyrin synthesis was detected in vivo [57]. Thus, these results suggest that β ankyrin also exhibits a pattern of decreased synthesis and concomitant stabilization in parallel with that of $\alpha\gamma$ spectrin during the later stages of neuronal differentiation.

The sharp decline in the level of $\alpha\gamma$-spectrin and β-ankyrin synthesis on embryonic day 15 coincides with a dramatic induction in the accumulation in the embryonic cerebellum of two new isoforms of ankyrin and spectrin, namely α ankyrin and β' spectrin. Immunofluorescence of developing cerebellum with β-spectrin antiserum has revealed that, throughout all stages of cerebellar ontogeny, mitotic cells and presumptive postmitotic granular cells that are migrating through the molecular layer and Purkinje cell layer exhibit no fluorescence [44]. Furthermore, there appears to be no β-spectrin fluorescence in the few presumptive granule cells that have migrated past the Purkinje cell layer prior to embryonic day 15/17 and which are forming the internal granular cell layer. The first indication of β-spectrin accumulation is on embryonic day 15/17, when small foci of β-spectrin fluorescence were detected at the plasma membrane of the perikarya of granule cells below the Purkinje cell layer. The degree of this fluorescence became more prominent by embryonic day 19, and by hatching (day 21) was found to be essentially indistinguishable from that in the adult cerebellum [43,44]. Before the onset of granule cell migration on embryonic day 15, the Purkinje cells have already migrated and lined up to form a distinct layer [40]. However, β-spectrin fluorescence was not detected in the perikarya of these cells until embryonic day 16/17, and approximately 24 hr before it was detected in the granule cells [44]. In addition, β-spectrin fluorescence was detected in the perikarya of the neurons of the cerebellar nuclei, but also only subsequent to embryonic day 16/17.

In the developing retina, we also detected the expression of $\alpha\gamma$ spectrin throughout neuronal differentiation in all developing layers and the developmentally regulated appearance of β' spectrin on embryonic day 17 at the plasma membrane of the perikarya and dendrites of ganglion cells and the perikarya of rods and cones [45].

Thus, in contrast to the constitutive expression of $\alpha\gamma$ spectrin and β ankyrin, the expression of β' spectrin and α goblin is developmentally regulated. This was demonstrated more directly by immunoautoradiography of SDS-solubilized extracts of day-13–21 embryonic chick cerebella [57] (see also Fig. 2D). On embryonic days 13–15, little or no β' spectrin or α ankyrin is detected. However, beginning on embryonic day 15/17, there is a dramatic induction in the accumulation of β' spectrin and α ankyrin, whose relative level of accumulation appears to reach a plateau by embryonic day 21 (hatching) [57]. As in the case of β ankyrin (see above), we were only able to determine qualitatively the levels of β' spectrin and α ankyrin synthesis in the developing retina. The results indicated that there was little or no synthesis of β' spectrin and α ankyrin in neurons on embryonic day 13 but detectable synthesis of both polypeptides by embryonic day 21, and also in the adult [57]. This suggests that, contrary to the decreasing levels of $\alpha\gamma$ spectrin and β ankyrin synthesis during the later stages of neuronal morphogenesis, there is an induction in those of β' spectrin and α ankyrin.

Two Successive Developmentally Regulated Programs of Protein Expression and Patterning on the Plasma Membrane are Involved in the Assembly of the Neuronal Membrane Skeleton

Our analyses of spectrin and ankyrin isoform expression and distribution in neurons during cerebellar and retinal ontogeny provide, for the first time, an insight into the sequence of molecular events involved in the temporal and spatial assembly of the membrane-skeleton during neuronal morphogenesis. We have hypothesized that there are two successive, developmentally regulated programs of protein expression and patterning on the plasma membrane during neuronal morphogenesis [57] (see Fig. 3). The first program commences upon withdrawal of the cells from the proliferative cell cycle, when $\alpha\gamma$ spectrin and β ankyrin gradually accumulate on the plasma membrane as a uniform pattern throughout the neuron during the phase of cell growth (Fig. 3). Upon synaptogenesis and the cessation of cell growth on embryonic day 15/17 [40], the levels of synthesis of $\alpha\gamma$ spectrin and β ankyrin decline rapidly but their steady state amounts continue to increase, indicating a down-regulation in the first program with a concurrent stabilization of newly synthesized proteins assembled onto the plasma membrane (Fig. 3). Concomitantly, there is an induction in the second program during the phase of terminal differentiation as manifested through the rapid induction in the synthesis and accumulation of $\alpha\beta'$ spectrin and α ankyrin in the perikarya and dendrites of neurons, which, together with $\alpha\gamma$ spectrin, continue to be synthesized in the adult (Fig. 3). This nonuniform pattern of assembly of α ankyrin and $\alpha\beta'$ spectrin is superimposed on the uniform pattern of β ankyrin and $\alpha\gamma$ spectrin assembled during the phase of cell growth, which together establish the asymmetry in the molecular composition of the neuronal membrane-skeleton.

These two patterns of ankyrin and spectrin isoform expression and patterning on the plasma membrane during the phases of cell growth and terminal differentiation appear to be representative of two general programs of structural protein expression

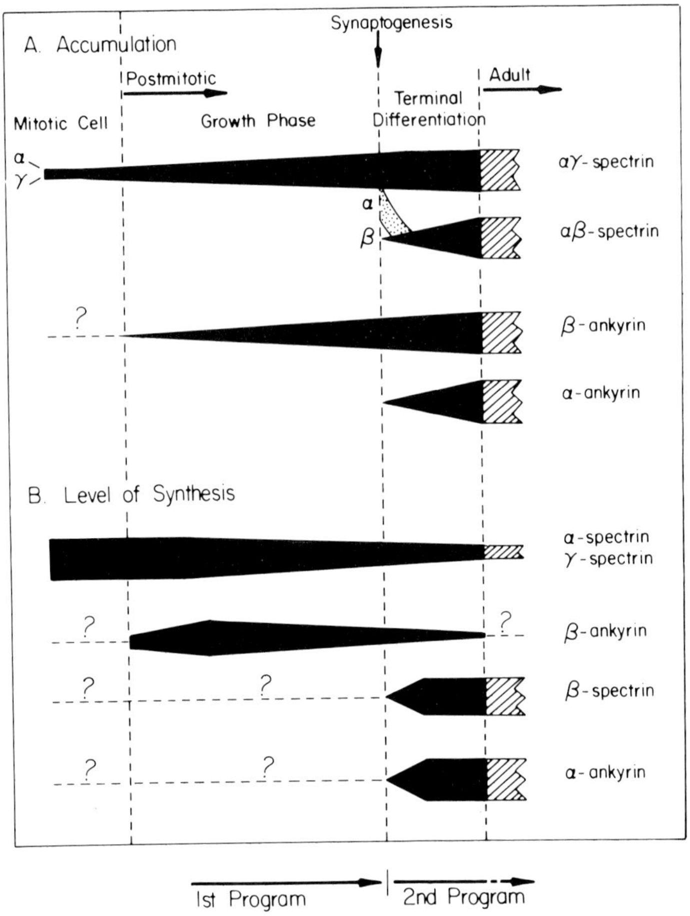

Fig. 3. Schematic representation of the developmentally regulated changes in the levels of synthesis and rates of accumulation of isoforms of ankyrin and spectrin during neuronal morphogenesis. The wedges are not drawn to scale and thus do not represent the relative amounts of each isoform at a given stage. The hatched areas indicate continued synthesis or presence at steady state of an ankyrin or spectrin isoform in the adult.

during neuronal morphogenesis. First, during the phase of cell growth, the gradual accumulation of β ankyrin and $\alpha\gamma$ spectrin is paralleled by an increase in the number of neurofilaments in direct proportion to the increase in cross-sectional area of the axon [13,27,39,77,78, see also 69]. At the same time, there is a decrease in the size of the soluble pools of tubulin and actin [68], presumably reflecting an increase in the number of microtubules and microfilaments in growing axons. Second, upon the onset of synaptogenesis and the cessation of cell growth, there is a sharp decline in the levels of tubulin and actin mRNA [15,68] and their levels of synthesis [68] as shown here also for β ankyrin and $\alpha\gamma$ spectrin, indicating a general down-regulation in the expression of proteins that may be involved in the outgrowth and stabilization of the axons and dendrites during the phase of cell growth. Concomitantly, during the phase of terminal differentiation there is an induction in the accumulation of a new set of structural proteins which includes MAP-2 [24,26,38,51] together with α ankyrin and $\alpha\beta'$ spectrin which are localized exclusively in the perikarya and dendrites, where they may modulate the topology of the plasma membrane or other cytoskeletal components in the region of the synapse (see below).

Localization of Specific Structural Proteins to the Perikaryon and Dendrites: Implications for Neuronal Function

The temporal and spatial appearance of α ankyrin and $\alpha\beta'$ spectrin has been shown to coincide remarkably with synaptogenesis in the neurons of the chick cerebellum and retina [44,45,57]. Several other proteins have also been shown to appear in conjunction with neuronal terminal differentiation, including Synapsin I, a synaptic vesicle phosphoprotein [25]; neuronal enolase [67], the glycoprotein Thy 1.1 [6], a 23,000-dalton neuronal mitochondrial-specific protein [36], and MAP-2 [38].

The functional significance of the appearance of α goblin and $\alpha\beta'$ spectrin in the perikarya and dendrites upon synaptogenesis and during subsequent neuronal terminal differentiation is at present unknown. However, since chicken neuronal α spectrin and β' spectrin are structurally homologous to their counterparts in erythrocytes (see above), the expression of $\alpha\beta'$ spectrin may be linked to the requirement of the neuronal plasma membrane to have a function analogous to that in the erythrocyte, namely, to maintain an invariant topological distribution of a discrete population of integral membrane proteins in the perikarya and dendrites, and perhaps in the region of the synapse, by restricting their mobility in the plane of the plasma membrane and thereby segregating them from axonal transport [45].

In addition, it may be significant that α ankyrin in neurons [57] and erythrocytes [7,2,3] and MAP-2 in neurons [72,74,76,79] are the target substrates for phosphorylation by cAMP- and Ca^{++} -dependent protein kinases. In neurons, cAMP and Ca^{++} are important second messengers which mediate, at least in part, the action of neurotransmitter signals that occur primarily on the plasma membrane of the dendrites and perikarya [for a review and references, see 35]. Hence, the specific phosphorylation of α ankyrin and MAP-2 in this way may produce dynamic changes in the membrane-skeleton and thus modulate the morphology of the plasma membrane of the dendrites or perikarya at the level of the synapse. Furthermore, in the avian erythrocyte, the hormone-dependent phosphorylation of goblin (ankyrin) has been shown to be correlated quantitatively with an increase in Na^{+} -K^{+} cotransport [2]. In view of the role of ionic conductance in the transmission of electrical impulses, the specific phosphorylation of ankyrin may play a regulatory role in this respect also in neurons.

ACTIVE MAINTENANCE OF THE ASYMMETRY IN THE NEURONAL MEMBRANE-SKELETON: IMPLICATION FOR AXONAL TRANSPORT

In the mature neuron, newly synthesized proteins and organelles that are destined for the presynaptic terminal are actively segregated from the perikaryon and transported down the axon. Several recent studies have shown that $\alpha\gamma$ spectrin is a major component of axonal transport [45,47–49], which, given the fact that $\alpha\beta'$ spectrin is localized exclusively in the perikaryon and dendrites, indicates that there is a mechanism for segregating these two forms of spectrin during the process of axonal transport.

Active Segregation of $\alpha\gamma$ Spectrin and $\alpha\beta'$ Spectrin in the Adult Chicken Optic System

The distribution of $\alpha\gamma$ and $\alpha\beta'$ spectrin in the adult chicken optic system has been determined by immunoautoradiography of the retina, optic nerve, and optic tract [45]. The results of this analysis graphically demonstrate the asymmetry in the distribution of these two forms of spectrin in the neuron. As expected from the immunofluorescence results, $\alpha\gamma$ spectrin was detected throughout the optic system, in the retina, optic nerve, and optic tract segments. However, $\alpha\beta'$ spectrin was localized exclusively in the retina, where we had shown previously that it is present predominantly in the perikarya of the ganglion cells (see above) [45].

To determine whether $\alpha\gamma$ and $\alpha\beta'$ spectrin are actively segregated from one another during the process of axonal transport, newly synthesized proteins in the retina were labeled radioactively with ^{35}S-methionine in vivo [45]. Ninety-six hours after labeling was initiated, the chicken was sacrificed and the optic system dissected. The distribution of ^{35}S-methionine-labeled $\alpha\gamma$ and $\alpha\beta'$ spectrin in the retina, optic nerve, and optic tract was then determined by immunoprecipitation and fluorography. The results demonstrated that $\alpha\gamma$ spectrin is present throughout the optic system, confirming that this form of spectrin is a component of axonal transport [45,47]. On the other hand, β' spectrin was detected exclusively in the retina and none was found in the optic nerve or optic tract. Therefore, $\alpha\beta'$ spectrin must be actively segregated from $\alpha\gamma$ spectrin and other components of axonal transport in the adult chicken optic system. Since we detected β' spectrin in only the perikarya of cerebellar neurons, whereas $\alpha\gamma$ spectrin was present throughout the neuron, we assume that $\alpha\beta'$ spectrin is also actively segregated from $\alpha\gamma$ spectrin in these neurons as well.

What is the nature of the mechanism(s) for segregating $\alpha\beta'$ spectrin from $\alpha\gamma$ spectrin and axonal transport, and confining it to the perikaryon? Previously, we hypothesized that one mechanism may be that there exists a high-affinity membrane receptor specific for β' spectrin which drives the stable assembly, and hence accumulation, of $\alpha\beta'$ spectrin in the perikaryon despite the presence of excess γ spectrin. The latter may bind to a different membrane receptor which is not retained in the perikaryon and thus $\alpha\gamma$ spectrin becomes a component of axonal transport. α Spectrin, by virtue of the fact that it can bind to either γ or β' spectrin, will assemble with either subunit in equimolar amounts [45]. This hypothesis was based upon two experimental results. First, comparison of the peptide maps of erythroid ($\alpha\beta$) spectrin and $\alpha\gamma$ spectrin [31,45,58] has shown that these two forms of spectrin differ in the polymorphic subunit (γ and β'/β spectrin) which, at least in erythrocytes, is known to be involved in linking the spectrin-actin complex to the plasma membrane; as noted earlier in the erythrocyte, β spectrin binds to an intermediary protein, ankyrin (goblin [56]; see below), which in turn is linked to a subset of the transmembrane-orientated anion transporters [for reviews see 8,16]. Thus the structural difference in the polymorphic subunits may result in the differential binding of the $\alpha\beta'$ and $\alpha\gamma$ spectrin complexes to specific membrane receptors.

Second, recent studies on the assembly of spectrin in embryonic chick erythroid cells [14,53] and in vitro reconstitution experiments with erythroid spectrin [52] have shown that the molar amount of newly synthesized α spectrin assembled onto the membrane is limited by the amount of β spectrin assembled. Furthermore, the assembly of the $\alpha\beta$-spectrin complex onto the membrane is limited by the availability of newly synthesized ankyrin (goblin) [53]. The binding of $\alpha\beta$ spectrin to ankyrin and to the membrane results in the stabilization of the assembled complex against proteolytic degradation, whereas unassembled subunits synthesized in excess of their membrane binding sites are degraded in the cytosol [14,53].

In view of the fact that ankyrin isoforms are present in adult chicken neurons, we sought to analyze the possibility that ankyrin plays a role in the active segregation of $\alpha\gamma$ and $\alpha\beta'$ spectrin.

Newly Synthesized α Ankyrin is Not a Component of Axonal Transport: A Possible Role for α Ankyrin in the Active Segregation of $\alpha\gamma$ and $\alpha\beta'$ Spectrin

Immunoautoradiography of the adult chicken optic system with ankyrin antiserum has demonstrated that, similarly to $\alpha\beta'$ and $\alpha\gamma$ spectrin, the α and β ankyrin isoforms have different distributions [57]. α Ankyrin is present exclusively in the retina together with $\alpha\beta'$ spectrin, whereas β ankyrin has a distribution throughout the optic system similar to that of $\alpha\gamma$ spectrin. This suggests that α ankyrin is segregated from axonal transport and is selectively retained together with $\alpha\beta'$ spectrin in the retina. Indeed, analysis of the spatial distribution of ^{35}S-methionine-labeled α ankyrin in the optic system 96 hr after labeling in situ has revealed that it is present almost exclusively in the retina and is absent from the optic nerve and optic tract. Thus, newly synthesized α ankyrin and $\alpha\beta'$ spectrin appear to be cosegregated from newly synthesized $\alpha\gamma$ spectrin and the components of axonal transport. Little or no synthesis of β ankyrin has been detected in the adult chicken retina [57].

Given the fact that α ankyrin and β' spectrin share extensive structural homologies with their counterparts in erythrocytes [45,57], one explanation of the cosegregation of α ankyrin and $\alpha\beta'$ spectrin may be that, as in the erythrocyte, α ankyrin acts as the specific membrane attachment protein for erythroid spectrin. In this respect, it is significant that the molar ratio of newly synthesized α ankyrin:β' spectrin in the retina labeled to steady state has been shown to be approximately 1:2 [57], which is similar to their molar ratio at steady state in chicken [56] and mammalian (for references see [8,16]) erythrocytes. This suggests that there is a relatively tight coupling between the levels of synthesis of α ankyrin and β' spectrin which may result in the rapid and complete saturation of newly synthesized α ankyrin binding sites with $\alpha\beta'$ spectrin and hence the efficient assembly of erythroid spectrin onto the plasma membrane of the perikarya and dendrites. Since we detected the synthesis of only α ankyrin in adult neurons, a consequence of the saturation of all newly synthesized α-ankyrin binding sites with $\alpha\beta'$ spectrin is that newly synthesized $\alpha\gamma$ spectrin would be excluded from the plasma membrane of the perikarya and dendrites and may thus become a component of axonal transport. An implicit assumption here is that α ankyrin has a higher affinity for erythroid spectrin than for brain spectrin. This is supported by the results of several studies which have shown that human erythroid ankyrin binds to erythroid spectrin with a significantly higher affinity than to brain spectrin [12,23].

Although the steady state distributions of $\alpha\gamma$ spectrin and β ankyrin in the adult optic system are also similar, we could detect little or no synthesis of β ankyrin in the adult retina, compared to $\alpha\gamma$ spectrin [57]. This suggests that β ankyrin is relatively resistant to proteolytic turnover and, in the absence of its synthesis in the adult, must have accumulated predominantly during the early stages of neuronal morphogenesis prior to hatching; as noted earlier, studies of the assembly of the erythroid membrane-skeleton have demonstrated that ankyrin assembled onto the plasma membrane is resistant to proteolytic degradation [53]. Thus, if β ankyrin is a membrane attachment protein for either $\alpha\beta'$ or $\alpha\gamma$ spectrin, it may perform this function only during the phase of cell growth, when it accumulates at steady state uniformly on the plasma membrane of postmitotic neurons. Since $\alpha\gamma$ spectrin is the predominant, if not exclusive, form of spectrin synthesized and assembled onto the plasma membrane during this stage of neuronal morphogenesis, this would result in the saturation of all available β-ankyrin binding sites on the axonal plasma membrane with $\alpha\gamma$ spectrin prior to the onset of accumulation of $\alpha\beta'$ spectrin and α ankyrin; consequently, in the adult axon there would not be any β ankyrin-mediated membrane binding sites for newly synthesized $\alpha\beta'$ or $\alpha\gamma$ spectrin. Additionally, it is possible that in the adult there is a protein(s) which acts as the specific membrane attachment site for $\alpha\gamma$ spectrin during axonal transport in the mature neuron but which is not recognized by the ankyrin antiserum used in these studies. In this respect, it may be significant that the kinetics of $\alpha\gamma$ spectrin movement down the axon show that it is a component of more than one group of transported proteins [47–49], indicating that $\alpha\gamma$ spectrin may have the capacity to bind to several different proteins other than α and β ankyrin and hence enter the axon as a component of axonal transport [57].

Are Other Components of the Erythroid Membrane-Skeleton Expressed in Neurons?

The demonstration that polypeptides structurally and functionally related to the erythroid membrane-skeleton proteins spectrin and ankyrin are expressed in neurons raises the possibility that further counterparts of erythroid-skeletal proteins may be present in these cells. Recently, two independent studies have addressed this question using antibodies raised against erythrocyte protein 4.1 with different results. Studies from this laboratory [34] did not detect immunoreactive forms of this protein in chicken cerebellum or retina by immunofluorescence or immunoautoradiography, although structurally similar proteins were unambiguously detected in both avian and mammalian erythrocytes and lens fiber cells. On the other hand, Goodman et al [32] have reported that a polypeptide serologically related to mammalian erythrocyte protein 4.1 is present in porcine cerebellum with a distribution similar to that of brain ($\alpha\gamma$) spectrin. At present, the reason(s) for these differences is not known and requires further investigation.

CONCLUSIONS

Our ongoing studies, reviewed here, have sought to gain an insight into the molecular basis of neuronal morphogenesis and the role of the membrane-skeleton in the establishment and maintenance of the structural and physiological asymmetry of the neuron. Using the components of the erythrocyte membrane-skeleton as a model system, we have demonstrated an asymmetry in the molecular composition of the neuronal plasma membrane which appears to parallel the characteristic structural asymmetry of the mature neuron; α ankyrin and $\alpha\beta'$ spectrin are detected exclusively in the perikarya and dendrites, whereas β ankyrin and $\alpha\gamma$ spectrin are present throughout the neuron. Furthermore, by analyzing the sequence of molecular events involved in the establishment of this asymmetry in the membrane-skeleton during neuronal morphogenesis, we have identified two discrete patterns of spectrin and ankyrin isoform expression and assembly on the plasma membrane during the phases of cell growth and terminal differentiation. When superimposed upon one another in the mature neuron, these two classes of spectrin and ankyrin isoforms may be responsible for establishing the structural asymmetry of the membrane-skeleton. At present, the details of the molecular mechanism(s) involved in maintaining this asymmetry in the membrane-skeleton are unknown. However, our analysis of the temporal and spatial accumulation in neurons of ankyrin and spectrin isoforms has demonstrated that this asymmetry is maintained actively in adult neurons. Furthermore, they have identified one of the ankyrin isoforms, α ankyrin, as a candidate for the membrane attachment site of $\alpha\beta'$ spectrin in the perikarya and dendrites, and hence the potential molecular basis for the active segregation of the two spectrin forms. This hypothesis is consistent with all the known biochemical properties of ankyrin in erythrocytes and neurons, and the kinetics of assembly of spectrin onto the plasma membrane of erythrocytes. However, conclusive evidence for the role of the various ankyrin isoforms in mediating the segregation of the two spectrin isoforms must await the purification of the individual polypeptides and their subsequent reconstitution in a well-defined in vitro system [52].

ACKNOWLEDGMENTS

The work reviewed here was supported by grants from the National Institutes of Health, the National Science Foundation, and the Muscular Dystrophy Association of America to E.L. W.J.N. was also supported by an international fellowship from the Cancer Research Campaign awarded by the International Union Against Cancer, a senior research followship from the American Heart Association Greater Los Angeles Affiliate, and by a grant from the National Eye Institute. E.L. is the recipient of a research career development award from the NIH.

REFERENCES

1. Addison WHF: J Comp Neurol 21:459–485, 1911.
2. Alper SL, Palfry HC, DeRiemer SA, Greengard P: J Biol Chem 255:11029–11039, 1980.
3. Alper SL, Beam KG, Greengard P: J Biol Chem 255:4864–4871, 1980.
4. Anderson RA, Lovrien RE: Nature 307:655–658, 1984.
5. Araki M, Maeda T, Kimura M: Brain Res Bull 10:97–102, 1983.
6. Barclay AN: J Neurochem 32:1249–1257, 1979.
7. Beam KG, Alper SL, Palade GE, Greengard P: J Cell Biol 83:1–15, 1979.
8. Bennett V: J Cell Biochem 18:49–65, 1982.
9. Bennett V, Stenbuck PJ: J Biol Chem 254:2533–2541, 1979.
10. Bennett V, Stenbuck PJ: Nature 280:469–473, 1979.
11. Bennett V, Stenbuck PJ: J Biol Chem 255:2540–2548, 1980.
12. Bennett V, Davis J, Fowler WA: Nature 299:126–131, 1982.
13. Berthold CH: "Physiology and Pathobiology of Axons." New York: Raven Press, 1978, pp 3–63.
14. Blikstad I, Nelson WJ, Moon RT, Lazarides E: Cell 32:1081–1091, 1983.
15. Bond JF, Farmer SR: Mol Cell Biol 3:1333–1342, 1983.
16. Branton D, Cohen CH, Tyler J: Cell 24:24–32, 1981.
17. Brenner SL, Korn ED: J Biol Chem 254:8620–8627, 1979.
18. Bunge MB, Bunge RP, Peterson ER: Brain Res 6:728–749, 1967.
19. Burridge K, Kelly T, Mangeat P: J Cell Biol 95:478–486, 1982.
20. Cantino D, Daneo LS: Experimentia 29:85–87, 1973.

21. Coulombre AJ: Am J Anat 96:153–190, 1955.
22. Daniels MP, Vogel Z: Brain Res 201:45–56, 1980.
23. Davis JQ, Bennett V: J Biol Chem 259:1874–1881, 1984.
24. DeCamilli P, Moretti M, Navone F, Lohman-Walter SM, Walter U: Abstr 13th Ann Meet Amer Soc Neurosci, 1983 (in press).
25. DeCamilli P, Cameron R, Greengard P: J Cell Biol 96:1337–1354, 1983.
26. DeCamilli P, Miller PE, Navone F, Theurkauf WE, Vallee RB: Neuroscience 1984 (in press).
27. Friede RL, Samorajski T: Anat Rec 167:379–387, 1970.
28. Glees P, Sheppard BL: Z Zellforsch Mikrosk Anat 62:356–362, 1964.
29. Glenney JR, Glenney P, Osborn M, Weber K: Cell 28:843–854, 1982.
30. Glenney JR, Glenney P, Weber K: J Biol Chem 257:9781–9787, 1982.
31. Glenney JR, Glenney P: Cell 34:503–512, 1983.
32. Goodman SR, Casoria LA, Coleman DG, Zagon S: Science 224:1433–1436.
33. Gozes I, Sweadner KJ: Nature 294: 447–480, 1982.
34. Granger BL, Lazarides E: Cell 37:595–607, 1984.
35. Greengard P: In "Intracellular Signals in the Brain." The Harvey Lecture, Series 25. New York: Academic Press, 1981, pp 277–331.
36. Hawkes R, Niday E, Matus A: Cell 28:253–258, 1982.
37. Hirokawa N, Glickman MA, Willard MB: J Cell Biol 98:1523–1536, 1984.
38. Huber G, Matus A: J Neurosci 4:151–160, 1984.
39. Hughes WF, LaVelle A: Anat Rec 179:297–302, 1974.
40. Jacobson M: In "Developmental Neurobiology." Second Edition. New York: Plenum Press, 1978, pp 75–92.
41. Kahn AJ: Dev Biol 38:30–40, 1974.
42. Lasek RJ: Neurosci Res Prog Bull 19:7–32, 1981.
43. Lazarides E, Nelson WJ: Science 220:1295–1296, 1983.
44. Lazarides E, Nelson WJ: Science 222:931–933, 1983.
45. Lazarides E, Nelson WJ, Kasamatsu T: Cell 36:269–278, 1984.
46. Leto TL, Marchesi VT: J Biol Chem 259:4603–4608, 1984.
47. Levine J, Willard M: Brain Res 194:137–154, 1980.
48. Levine J, Willard M: J Cell Biol 90:631–643, 1981.
49. Lorenz T, Willard M: Proc Natl Acad Sci USA 75:505–509, 1978.
50. Lux SE, John KM, Karnovsky MJ: J Clin Invest 58:955–963, 1976.
51. Maseck A, Fellows A, Francon J, Nunez J: Nature 284:353–355, 1980.
52. Moon RT, Lazarides E: Nature 305:62–65, 1983.
53. Moon RT, Lazarides E: J Cell Biol 98:1899–1904, 1984.
54. Mugnaini E: In Llinas R (ed): "Proceedings of the First International Symposium of the Institute for Biomedical Research: Neurobiology of Cerebellar Evolution and Development." Chicago: Institute for Biomedical Research, 1969, p 794.
55. Nelson WJ, Lazarides E: Proc Natl Acad Sci USA 80:363–367, 1983.
56. Nelson WJ, Lazarides E: Proc Natl Acad Sci USA 81:3292–3296, 1984.
57. Nelson WJ, Lazarides E: Cell 39:309–320, 1984.
58. Nelson WJ, Granger BL, Lazarides E. J Cell Biol 97:1271–1276, 1983.
59. Nelson WJ, Colaco CALS, Lazarides E: Proc Natl Acad Sci USA 80:1626–1630, 1983.
60. Nishimura Y: Anat Embryol 158:329–347, 1980.
61. Oppenheim RW, Foelix RF: Nature 235:126–128, 1972.
62. Palfrey MC, Schieber W, Greengard P: Proc Natl Acad Sci USA 79:3780–3784, 1982.
63. Rager G: Proc R Soc Lond [Biol] 192:331–352, 1976.
64. Repasky EA, Granger BL, Lazarides E: Cell 29:821–833, 1982.
65. Rees RP, Bunge MB, Bunge RP: J Cell Biol 68:240–263, 1976.
66. Saetersdal TAS: Univ Bergen Arbok Naturvitenskap Rekke 3:1–53, 1956.
67. Schmechel DE, Brightman MW, Maranges PJ: Brain Res 190:195–214, 1980.
68. Schmitt M, Gozes I, Littauer UZ: Brain Res 121:327–342, 1977.
69. Shaw G, Weber K: Nature 298:277–279, 1982.
70. Sheffield JB, Fischman DA: Z Zellforsch 104:405–418, 1970.
71. Shimono T, Nosaka S, Sasaki K: Brain Res 108:279–294, 1976.
72. Sloboda RD, Rudolph SA, Rosenbaum JL, Greengard P: Proc Natl Acad Sci USA 72:177–181, 1975.

73. Steizner D, Martin AM, Scott GL: Z Zellforsch Mikrosk Anat 138:475–488, 1973.
74. Therkauf WE, Vallee RB: J Biol Chem 257:3284–3290, 1982.
75. Tyler J, Hargreaves W, Branton D: J Biol Chem 255:7034–7039, 1979.
76. Vallee RB: Proc Natl Acad Sci USA 77:3206–3210, 1980.
77. Weiss PA, Mayr R: Proc Natl Acad Sci USA 68:846–850, 1971.
78. Willard M, Simon C: Cell 35:551–559, 1983.
79. Yamauchi T, Fujisawa H: Biochem Biophys Res Commun 109:975–981, 1982.

Neurobiology: Molecular Biological Approaches to Understanding Neuronal Function and Development 93–100 (1985)

Lineage-Based Control of Neuronal Gene Expression

J. Gregor Sutcliffe, Robert J. Milner, Joel M. Gottesfeld, and Wanda Reynolds

Department of Molecular Biology, Research Institute of Scripps Clinic, La Jolla, California 92037

A genetic element called an ID sequence is located in the introns of most neuronal-specific genes. It is transcribed in neurons, but not in other tissues, by RNA polymerase III, and naturally appears in the RNA polymerase II primary transcripts of brain but not other tissues. We discuss a model in which the ID sequences act as cis regulators of neuronal-specific transcription and are positively activated by neuronal-specific transcription factors. Structural peculiarities of ID elements prevent nucleosomes from masking this DNA from solution reactions. Our model implicates the transcription specificity factors as the primary phenotypic manifestations of differentiation. We argue that ID sequences are the elements that inform a gene of the lineage of the cell.

Key words: brain transcription, gene regulation, promoters

What formally differentiates higher eukaryotes from prokaryotes is that the cells of higher eukaryotes must first differentiate into organs, each composed of one or more specific cell types, and then those organs must respond to ambient stimuli reflecting the physiological state of the whole organism. In order to achieve a differentiated state, batteries of genes must be activated in response to certain stimuli, and the required responses to the same stimulus may be different for various cells. The result of this process is that different genes are available for activation in different organs. This implies, a priori, two distinct levels of gene control, one related to the differentiation state and the other responding to ambient signals. These two levels of necessary gene regulation are known to developmental biologists, and their phenotypic manifestations are classically referred to as cell determination and cell position. We should expect that individual genes might contain sequences that detail a cell's lineage history and also sequences that can respond to ambient signals. Therefore, whereas the important cis acting sequences governing the control of prokaryotic genes cluster in small regions very near the initiation site for gene transcription, one expects that more extensive regions of cis control may affect eukaryotic transcription. Many recent studies have shown that certain regulators may act at considerable nucleotide distance from promoters.

Received April 1, 1984.

NEURONAL LINEAGE ELEMENTS

The studies that led to defining a neuronal-specific pre-mRNA sequence were motivated by an intent to define neuronal-specific proteins. Since mRNAs and proteins are likely to be an equivalent set for any organ, we constructed cDNA clones of rat brain mRNAs [1]. We analyzed these clones by RNA (Northern) blot analysis, comparing the hybridization of some 190 individual nick-translated cDNA clones to the poly A^+ RNA from brain, liver, and kidney, thus learning the size, tissue distribution, and relative abundances of a cross-section of brain mRNAs. We found that 56% of the brain mRNA mass (some 30,000 distinct, mostly rare mRNA species) was exclusively expressed in the brain [1]. Another 26% of the mRNA mass (at least 700 distinct mRNAs) represent species shared by the three test tissues, which are nonetheless highly regulated. The remaining 18% of the molecules (at least 600 species) are constitutively expressed in all three test tissues. Three percent of our clones hybridized to a brain-specific 160 nucleotide poly A^+ RNA species, even though the clones themselves had much larger (561–1225 bp) cDNA inserts [2]. These clones share an 82-nucleotide sequence that we refer to as an identifier (ID) sequence. The small RNA (which we call BC1) is absent from adrenal, spleen, testes, heart, muscle, intestine, lung, thymus, liver, and kidney. It is present in brain, pituitary, and solar plexus (a peripheral nervous organ) [3]. We believe that BC1 is a neural marker because it is absent from glial tumor lines, and the relative distribution of BC1 is inversely related to the glial distribution in the brain.

BC1 RNA is almost exclusively cytoplasmic (about 2,000 molecules per cell) and co-migrates with 5.8S RNA. A second RNA species, BC2, migrates as a diffuse smear of 100–110 nucleotides, hybridizes with the same probes, and seems to share all other properties with BC1 except that it seems to have a lower affinity for oligo d(T) cellulose. In highly purified brain poly A^+ RNA preparations, BC1 is clearly visible on stained gels. We have conducted a developmental study, isolating poly A^+ RNA from the brains and also from peripheral organs of rats at days 14, 16, 18, and 20 of embryonic development (day 14 is whole head) and 1, 5, 10, 15, 20, 25, and 30 days after birth. BC1 and BC2 RNAs are detected in hybridization experiments in large amounts beginning at 5–10 days postnatal. Our ID probes hybridize with considerably lesser intensity to all earlier samples, even though probes for control mRNAs hybridize well with the early samples. Work by several groups [4] has shown that neuronal chromatin in the mammals exhibits a shorter nucleosome repeat than either glial or other non-neuronal tissues. The onset of this change is between 5–10 days post natal in rats [5], the same time we see a large increase in BC1 and BC2 transcription. A few days thereafter, synthesis of adult brain proteins begins including that for the 1B236 neurotransmitter precursor whose mRNA we have cloned [6]. For 1B236, both mRNA transcription and protein synthesis begin between postnatal days 10 and 15.

We have sequenced four of the cDNA clones that hybridize to BC1 and BC2 and find that the clones contain a common core sequence of 82 nucleotides followed by an A-rich tract that we have called an identifier (ID) sequence [2,7]. The remaining cDNA sequences are dissimilar, except that they exhibit some of the characteristics of introns or other non-coding regions such as strings of repeated short oligonucleotides, long tracts of purines or pyrimidines, or homopolymeric tracts. A comparison of the regions of homology from the four cDNA clones has been made with four other homologous sequences found in an intron of rat growth hormone, a non-coding region of Harvey Sarcoma Virus (just 5′ to the ras gene), a rat pseudo-tubulin gene, and a rat U1-RNA gene. The comparison shows that the ID sequences are remarkably well conserved: 62 nucleotides are invariant, 15 positions differ in only one of the eight isolates, two positions differ in two of the isolates, and the other three positions are exclusively either purine or pyrimidine [7]. This is a much higher degree of conservation of repetitive genomic sequences than has been observed for previously described elements such as Alu sequences.

We have mapped the non-homologous regions of the ID clones by Northern blotting. Two of these hybridize to brain-specific RNAs, and a reasonable case can be made that the other two may be related to brain-specific mRNAs as well [7]. A restriction fragment from one of our ID-containing cDNA clones hybridizes to a brain-specific mRNA, but its adjacent fragment does not hybridize to any target. Since both fragments hybridize to the same Southern blot target, this suggests that the hybridizing fragment corresponds to an exon of a brain specific gene and the adjacent non-hybridizing fragment corresponds to an intron [7]. The nucleotide sequence supports this suggestion, since the hybridizing fragment contains an open reading frame whereas the non-hybridizing fragment does not. Thus we believe that the ID clones are derived by copying internal regions of hnRNA precursors of mRNA that appear as a contaminant of our cytoplasmic RNA preparation. They are likely to represent cDNA copies of RNA polymerase II (Pol II) transcripts.

Nonetheless, the highly conserved ID sequence shows homology with the consensus RNA polymerase III transcription initiation sequence. In fact, cDNA clones containing ID sequences function as Pol III transcription units in vitro in HeLa S100 extracts. By cleaving the clones with a variety of restriction endonucleases, we have found that Pol III initiation begins at position 1 of the core sequence of 82 nucleotides [3]. This corresponds exactly to the 5′ ends of the in vivo BC1 and BC2 RNAs, as shown by primer extension by reserve transcriptase. Correct initiation of Pol III is known to be dependent on two cis acting sequences (A block and B block), which lie within Pol III transcription units; that is, 3′ to the RNA start site [8]. Several trans acting factors (other than the polymerase itself) are known to be involved in Pol III transcription of tRNA and 5S genes.

In order to resolve the issue of whether ID sequences are transcribed in vivo by Pol II, Pol III, or both, we have used nuclei from brain, liver, and kidney cells to program transcription in a HeLa S100 system [9] of ^{32}P-labelled RNA in vitro and hybridized the transcription products to filters on which ID-containing plasmid DNAs were bound [3]. After correction for hybridization efficiency, we find that 62% of the brain transcripts—but only background levels (less than 4%) of liver or kidney transcripts—hybridize to the filter and thus contain ID sequences. Most of the incorporation in brain is in neuronal cells, even though glial cells represent most of the mass. When the filter-bound RNA is eluted and examined by gel electrophoresis, two sorts of products are observed: large heterogeneous RNA and a group of small 100–110 nucleotide species. The synthesis of the large molecules is sensitive to α-amanitin (at 1 μg/ml, a concentration that inhibits Pol II but not Pol III transcription), whereas the synthesis of the smaller species, which are probably the BC2 species observed in vivo, is resistant to α-amanitin. Therefore, ID sequences are located in a substantial portion of brain transcripts synthesized by both Pol II and Pol III but are missing from liver and kidney transcripts. Interestingly, liver and kidney program a distinct set of 100–110 nucleotide RNAs, which do not hybridize with the ID sequence and the synthesis of which is resistant to α-amanitin, suggesting analogous Pol III transcription elements in those tissues. All of these studies give qualitatively similar results whether the transcription is conducted in HeLa S100 extracts [9], in HeLa whole-cell extracts [10], or with no added protein (endogenous synthesis).

If liver chromatin is treated with 0.35 M NaCl so that the nucleosomes are allowed to slide, transcription of RNA molecules containing the ID sequences occurs, but only the large heterogeneous species are seen. This is not true for NaCl-treated brain chromatin, where no change in the pattern is observed, although the incorporation is considerably decreased. This suggests that nucleosomes are repressing Pol II synthesis of RNA-containing ID sequences in the liver, and that the transcription by Pol III of the small RNA, which we think is BC2, requires neuronal-specific transcription factors [3].

Dr. William Hahn's laboratory at the University of Colorado has used our ID clones to probe preparations of nuclear RNA made from brain, liver, and kidney. Dr. Hahn finds that both large poly A^+ and poly A^- nuclear RNA from brain hybridize strongly to the ID probes in Northern blot experiments. The probes do not hybridize detectably with the liver or kidney poly A^+ or poly A^- nuclear RNAs (personal communication). Thus, ID sequences are found in both Pol II (hnRNA) and Pol III (BC2) transcripts made only in neural tissue. BC1 might be a polyadenylated version of BC2 (which possibility we will discuss later). Hahn and his colleagues have also described a class of poly A^- RNA in rodent brain with equivalent but non-overlapping complexity to the class of poly A^+ RNAs that appears postnatally [11]. It is not known whether the brain-specific, poly A^- RNAs are made by Pol II or Pol III, but one possible explanation for their origin is that their synthesis is initiated at ID sequences by Pol III and ends at Pol III terminators; hence they are not generally polyadenylated. The fact that a majority of the brain poly A^- species terminate in U (as do most Pol III transcripts) is consistent with this explanation. Although this may explain the origin of brain poly A^- RNA, it does not address the possible role of these species in encoding brain-specific proteins. It does, however, explain their postnatal appearance, since BC1 and BC2 transcription largely begins after 5 days postnatal development.

Given the abundance of ID sequences in brain-specific transcripts, there must be many copies in the rat genome. Indeed, measurements utilizing a phage library of rat genomic clones with inserts of, on average, 15 kilobases show that 43% of the clones contain at least one ID sequence. This means that there are 0.9–1.5 $\times$ 10^5 ID copies per genome, a number in excess of the 30,000 adult brain-specific poly A^+ genes [7]. The transcription experiments suggest that ID sequences are specifically linked to brain genes, since they are found in brain but not liver or kidney primary transcripts. However, since only a small percentage (3%) of brain cDNA clones and brain mRNAs contain ID sequences, but about 60% of the primary transcripts contain the ID sequence, the linkage in primary transcripts must be unstable. That is, the ID sequences are located in the introns of most adult neuronal-specific poly A^+ genes (at least by mass). From the phage genomic library, we isolated the gene corresponding to the 1B236 mRNA that encodes a neuronal-specific precursor of a neurotransmitter [6] and used fragments from the 1B236 cDNA clone to map intron and exon regions of the gene. The 1B236 gene contains at least seven introns, of which two contain ID sequences [7]. Thus, it seems that ID sequences are highly correlated with neuronal-specific genes and occupy some, but not all, introns of at least most adult neuronal-specific genes.

What portion of ID sequences are active in a given neuron? We know from our studies [1,6] that most of the 30,000 brain-specific poly A^+ mRNAs are individually expressed in only a small subset of neurons. What is the state of the ID sequences in their genes? We have isolated nuclei from brain, liver, and kidney and digested their chromatin with various concentrations of DNase for fixed times, then extracted the DNA from the samples, denatured it, and bound it in an array to a nitrocellulose filter by "slot blotting." The filter was then hybridized with an ID probe. The amount of hybridization to the liver and kidney samples was unchanged by the DNase treatment. However, with increasing concentrations of DNase, the hybridization to the brain DNA decreased to less than 40% of the untreated value without reaching plateau. This result apparently indicates that all ID sequences are accessible to DNase in all neurons, even though the genes in which they are located are transcribed by Pol II into hnRNA products in only small subsets of neurons.

MODEL FOR NEURONAL GENE CONTROL

The model that we have proposed to explain these data is that ID sequences are located within introns or adjacent to the 30,000 adult neuronal genes and act in cis. A factor present only in adult neurons acts in trans to direct transcription of the ID sequence by Pol III, and this event activates the neuronal genes in a primary manner, probably by altering chromatin structure. This gene activation is necessary but not sufficient to initiate Pol II transcription. A second control level must exist, because genes such as that encoding the 1B236 neurotransmitter protein are only active in certain brain cells. The second level of control may operate at promoter sequences in response to ambient signals. ID sequences may be the neuronal lineage-specific control sequences. Although circumstantial, their correlation with neuronal-specific genes is compelling. The observation that ID sequences are transcribed by Pol III in vitro as plasmid DNA and in brain nuclear extracts and that there may be brain-specific ID transcription factors suggests, but does not prove, that Pol III acting at the ID sequences might be involved in neuronal-specific Pol II initiation. We believe that BC2 is the primary Pol III transcript of the brain ID sequence and that BC1 could be generated by occasional polyadenylation of the Pol III transcripts. BC1 might be functional in its own right, performing some neuron-specific role, but it is not a necessary player in our model. Polyadenylation of BC2 producing BC1 could occur if some of the 10^5 ID sequences in the genome have mutated such that a T appears in their oligo(A) 3′ end, producing the AATAAA signal for cleavage and polyadenylation that is known to work on Pol II transcripts. Some other tissues probably have an analogous ID control system, because liver and kidney contain transcripts like BC2 that have different sequences. We have not observed a poly A^+ RNA of the BC1 size class in gel analysis of preparations of liver or kidney RNAs, but the liver and kidney nuclei do make BC2-sized Pol III transcripts [3].

ID ELEMENTS HAVE A PECULIAR DNA STRUCTURE

We discussed above that ID sequences show homology to (and act as) Pol III promoters and have the customary A-block, B-block structure. The A-block of ID is homologous in part to nucleotides 42–48 of the Xenopus 5S gene. This region of the 5S gene is just within the binding site for the 5S specific transcription factor, TFIIIA

[12], and, as naked DNA, the region is preferentially digested by the single-strand-specific S1 nuclease [13]. It may be no coincidence that TFIIIA, when bound, induces a DNase I hypersensitive site at approximately this position (+50). The correlation between an S1 site in naked DNA and a TFIII-induced DNase I site suggests a DNA perturbation from B-form that could somehow be involved in factor recognition.

These studies indicate an unusual DNA conformation within the A-block region of 5S promoter DNA. There are a number of unusual DNA conformations that are known to produce regions of single-stranded character in plasmid DNA. Such conformations include stem-loop formations or cruciforms, in which the unpaired terminal loop is preferentially cleaved [14]. Also, sequences of low melting temperature, such as AT-rich regions, are more prone to breathing. Another example is Z-DNA, in which the B–Z junctions are unstable and accessible to S1 [15]. Nordheim and Rich have attempted to correlate S1-sensitive sites within the SV40 enhancer with short tracts of Z-forming DNA (alternating purine-pyrimidine). Hypersensitive regions, however, do not always fit neatly into one of these categories. It is probable that certain sequences for reasons not readily apparent, are unstable when under torsional stress. Regions of instability recognized by nucleases are likely to be similarly apparent to nucleosomes and transcription factors or other elements of the transcriptional machinery.

Since ID sequences contain Pol III A-blocks, this leads us to suggest a mechanistic version of the formal model of ID control of Pol II genes described above.

MECHANISM OF ID FUNCTION

It is possible that genes are repressed by nucleosomes for Pol II transcription unless those genes contains a tissue-specific ID sequence that has been activated by that tissue's particular factors (such as Pol III transcription factors). Unless ID activation occurs, Pol II promoters cannot be recognized by their own trans activators. We must, therefore, explain how it is that ID sequences can be recognized by proteins, while Pol II promoters cannot. Peculiar DNA structure may be the key. In fact, it could solve two problems at once. We must assume that whatever the differentiation program is, it provides an ordered series of trans activitors of ID-type sequences. These positive regulatory factors must be able to interact with their target ID sequences even though most DNA is shrouded with histones. It must be that the various tissue ID sequences are not inhibited by chromatin structure from protein interactions. We suggest that this is because of the peculiar DNA conformation assumed by ID sequences, which interacts differently with nucleosomes than does other DNA. The difference may or may not be enough to exclude nucleosomes, but it must leave ID DNA available for solution reactions. We suggest that it also might order, either actively or passively, nucleosome binding in a manner that will be operationally defined in a moment. Thus, whenever a new stage of differentiation is achieved by a cell, new factors are produced that can interact with their cognate ID sequences. We postulate that some aspect of this interaction, possibly activation of Pol III transcription, affects local nucleosome interactions, disrupting the solenoid repeating structure of chromatin (perhaps by cooperative allosteric changes of adjacent nucleosomes), thus sending a signal over several kilobases that renders that DNA accessible to protein interaction. It may be that nucleosome ordering due to the odd DNA confor-

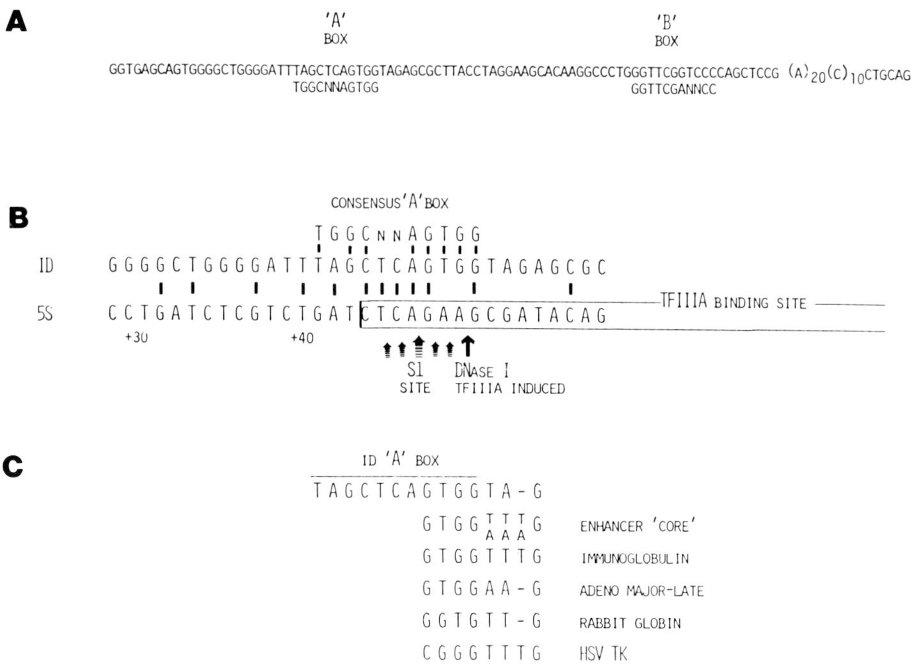

Fig. 1. Sequence homology between ID, the Pol III promoter, and enhancer elements. A) Nucleotide sequence of the ID region of plasmid p2A120 [2]. The consensus sequences for the Pol III A and B boxes are shown below. B) The region of ID containing the Pol III A box is homologous to a region of the 5S gene just within the TFIIIA binding site (+44 to +95). This region of the 5S gene is hypersensitive to S1 digestion (shaded arrows) as naked supercoiled plasmid DNA and is similar (if not identical) to the TFIIIA-induced DNase I hypersensitive site (solid arrow). C) The ID A box contains a sequence that is closely homologous to the consensus core-enhancer sequence derived by Weiher [16]. This sequence is present at three positions within the immunoglobulin enhancer [17]. Sequences with close homology have been found upstream of certain Pol II promoters, including Adenovirus major-late (−60 to −70) [18], rabbit β-globin (−75 to −85) [19], and TSV-TK [20].

mation at ID sequences allows such a signal to be propagated over a domain, but that the organization of nucleosomes around ID sequences in neighboring domains limits the extent of influence of individual ID elements. Such spheres of influence might correspond to what has been referred to as active and inactive domains of chromatin.

One consequence of this logic is that even genes that are constitutively expressed in all tissues will require ID sequences. These ID sequences will represent some degenerate, general form that reacts with a factor present in all cell types. Viral enhancers could be representative of such sequences, acting as the "lineage" markers for non-tissue-exclusive genes. They contain regions of Z-DNA, which may alter the way in which nucleosomes interact with the DNA and also order their local structure. The SV40 enhancer sequence is not covered by nucleosomes. If enhancers are a subclass of ID sequences—since it seems unlikely that they are transcribed by Pol III—it may be that specific protein interactions at brain ID sequences are more important than the production of a transcript. However, it may be that important features of Pol III promoters are the same features required for proper ID function.

The sequence $GTGG^{TTT}_{AAA}G$ or a closely related sequence is present in all viral enhancers and is essential for enhancer function [16]. Three copies of this sequence are present within the 313 bp restriction fragment of the Ig intron, which has been shown to contain the enhancer activity [17]. Similar sequences are also present within the regulatory elements upstream from certain Pol II promoters, including Adenovirus major late [18], rabbit β-globin [19], and HSV-TK [20]. The ID sequence contains a closely homologous sequence, GTGGTAG, in a position overlapping the Pol III A-block (shown in Fig. 1). This may be the site at which the tissue-specific trans activators interact with ID sequences. Dissection of the brain ID sequence and comparison with the ID sequences of other tissues should be informative.

Although our evidence for a lineage-based control system for tissue-specific genes is derived from the study of the transcripts of one tissue, the brain, it seems logical that a similar system works for other tissues. We expect that most genes, tissue-specific or general, will contain cis regulators that determine whether they occupy "active regions" of chromatin in each particular cell type. Some of these will be Pol III promoters, as we expect to find in liver and kidney, but others may act through different mechanisms. Enhancers may fall into this class of element. The trans acting factors that govern the function of these cis acting sequences must be the primary phenotypic markers of lineage differentiation. The control of the expression of these factors *is* differentiation.

ACKNOWLEDGMENTS

We thank Linda Eddy for preparing the manuscript. This work was supported in part by a grant from the National Institutes of Health (GM32355) and McNeil Laboratories. This is publication No. 3441–MB from the Research Institute of Scripps Clinic.

REFERENCES

1. Milner RJ, Sutcliffe JG: Nucleic Acids Res 11:5497, 1983.
2. Sutcliffe JG, Milner R, Bloom FE, Lerner RA: PNAS 79:4942, 1982.
3. Sutcliffe JG, Milner RJ, Gottesfeld JM, Lerner RA: Nature 308:237, 1984.
4. Thomas JO, Thompson RJ: Cell 10:633, 1977.
5. Greenwood PD, Brown IR: Neurochem Res 7:965, 1982.
6. Sutcliffe JG, Milner RJ, Shinnick TM, Bloom FE: Cell 33:671, 1983.
7. Milner RJ, Bloom FE, Lai C, Lerner RA, Sutcliffe JG: PNAS, 81:713, 1984.
8. Hofstetter H, Kressmann A, Birnsteil ML: Cell 24:573, 1981.
9. Weil PA, Segall J, Harris B, Ng S-Y, Roeder RG: J Biol Chem 254:6163, 1979.
10. Manley JL, Fire A, Cano A, Sharp PA, Gefter ML: 77:3855, 1980.
11. Chandhari N, Hahn WE: Science 220:924, 1983.
12. Engleke DR, Ng S-Y, Shastry BS, Roeder RG: Cell 19:717, 1980.
13. Reynolds W, Gottesfeld J: (submitted for publication).
14. Singleton C, Klypik J, Stirdiwant S, Wells R: Nature 299:312, 1982.
15. Nordheim A, Rich A: In Gluzman Y (ed): "Enhancers and Eukaryotic Gene Expression," Cold Spring Harbor, New York: Cold Spring Harbor Laboratory, 1983, p 45.
16. Weiher H, Konig M, Gruss P: Science 219:626, 1983.
17. Banerji J, Olson L, Schaffner W: In Gluzman Y (ed): "Enhancers and Eukaryotic Gene Expression," Cold Spring Harbor, New York: Cold Spring Harbor Laboratory, 1983, p 136.
18. Hen R, Sassone-Corsi P, Corden J, Gaub MP, Chambon P: 79:7132, 1982.
19. Grosveld GC, Rosenthal A, Flavell RA: Nucleic Acids Res 10:4951, 1982.
20. McKnight S, Kingsbury R: Science 217:316, 1982.

Neurobiology: Molecular Biological Approaches to
Understanding Neuronal Function and Development 101–111 (1985)

Developmental Studies of Phospholipid Ca^{2+}-Dependent and S-100-Modulated Protein Phosphorylation Systems in Brain

R. Scott Turner, De-Fang Qi, and J.F. Kuo

Department of Pharmacology, Emory University School of Medicine, Atlanta, Georgia 30322

Developmental changes in the protein phosphorylation/dephosphorylation systems in rat brain were investigated. It was found that the activity level of phospholipid-sensitive Ca^{2+}-dependent protein kinase (PL-Ca-PK) in the particulate fraction of grey and white matter and the soluble fraction of grey matter increased rapidly and markedly after birth, reached the highest level at day 30, and declined slightly or remained unchanged thereafter. The enzyme level in the soluble fraction of white matter, in contrast, remained constant throughout the development and maturation of brain. Various ontogenetic changes in the substrate proteins for PL-Ca-PK were also noted. The levels of myelin basic protein (MBP) and other substrates (notably the 87K, 58K, 54K, and 50K M_r proteins in grey matter) progressively increased during development, reaching the highest at adulthood. These proteins, termed "adult proteins," may be involved in the maturation or aging of brain. The level of the 66K M_r protein from the particulate fraction of white and grey matter, on the other hand, increased rapidly after birth, reached a peak at day 18, and then declined to the initial neonatal level at adulthood. This protein, a major substrate for the enzyme and termed "neonatal protein," may be crucial for brain development. It is concluded that phospholipid/Ca^{2+}-dependent protein phosphorylation reaction and the sequential replacement of certain substrate proteins are involved in the development and maturation of the brain.

The developmental aspect of a newly-found protein phosphorylation system in the brain modulated by S-100 protein was also investigated. The phosphorylation of an endogenous substrate protein (M_r = 19K) for the S-100-modulated enzyme (tentatively referred to as protein kinase X) was stimulated by S-100, but not by calmodulin, in a Ca^{2+}-independent manner. The substrate protein was present at a much higher level in brain from adult than neonatal (2-day old) rats, a developmental change similar to that seen for protein kinase X. It is suggested that a protein phosphorylation system modulated by S-100 may be involved in regulation of certain mature neural functions as well.

De-Fang Qi's present address is the Shanghai Institute of Biochemistry, Chinese Academy of Sciences, Shanghai, People's Republic of China.

Received April 1, 1984.

The activity levels of phosphoprotein phosphatases (assayed using ^{32}P-labeled MBP, histone, and protamine-SO_4 were found to increase or decrease only slightly (up to 60%) in certain fractions from different brain regions during development, suggesting that phosphorylation, but not dephosphorylation, may be the rate-limiting step in determining the phosphorylation state of brain proteins.

Key words: **brain development, phospholipid, calcium, S-100, protein phosphorylation, myelin basic protein**

Since the discovery [1] of phospholipid-sensitive Ca^{2+}-dependent protein kinase (PL-Ca-PK, protein kinase C), the enzyme has been purified to near homogeneity from bovine heart [2] and to apparent homogeneity from rat brain [3] and pig spleen [4], and its properties have been characterized [2–4]. Endogenous substrate proteins specific for PL-Ca-PK have been observed in many tissues [5–9]. Subsequently, we have identified myelin basic protein (MBP) in brain [10,11] and troponin I and T subunits in cardiac [12] and skeletal [13] muscle to be substrates for PL-Ca-PK. The enzyme, its substrates, and its regulation have been reviewed elsewhere [14–16].

We have reported earlier that the activity levels of PL-Ca-PK in the soluble fraction of guinea pig brain are lower in the fetus than in the adult [17]. The ontogenetic changes in the enzyme levels in many other tissues are similar to, or different from, the change seen for brain [17]. We extend the preliminary observations made in guinea pig brain to the soluble and particulate fractions of grey and white matter of rat brain, studying the developmental changes in the activity level of PL-Ca-PK, as well as the changes in substrate proteins for protein phosphorylation systems stimulated by phospholipid/Ca^{2+} or calmodulin/Ca^{2+}.

Since the discovery of S-100 almost 20 years ago [18] and subsequent studies by many investigators, the biological roles of S-100 remain unknown. It has been speculated that S-100 may be involved in ion movements in membranes, genomic regulation in neural tissues, or the process of learning [19,20]. S-100, like calmodulin, causes a Ca^{2+}-induced disassembly of brain microtubules in vitro [21]. To date, there is no other evidence showing that S-100 could modify activities or properties of any other biological preparations, including enzymes. We found the presence in brain of a S-100-modulated protein kinase (tentatively referred to as protein kinase X), its endogenous substrate protein (M_r = 19K), and their developmental changes in brain.

Because phosphoprotein phosphatases, in addition to protein kinases, are responsible for the phosphorylation states of endogenous proteins, we have also determined the developmental changes of the phosphatase activity levels in the same studies.

MATERIALS AND METHODS

PL-Ca-PK was partially purified from pig brain (about 10% homogeneous) through the step of Sephacryl S-100, as described for the spleen enzyme [4]. Assay conditions for PL-Ca-PK were as described elsewhere [2,4,9,10,22] and in Figure 1. Protein kinase X was partially purified from pig brain as described and assayed based upon the ability of S-100 to promote phosphorylation of protamine (free base) [23]. PL-Ca-PK or protein kinase X activity was linear with respect to time of incubation and amount of enzyme. Affinity chromatography, using S-100-coupled Sepharose 4B, of endogenous substrate proteins for protein kinase X was as described [24]. The methods of phosphorylation of endogenous substrate proteins from brain and subsequent electrophoresis and autoradiography were as reported [6]. ^{32}P-proteins were prepared as described [25], and the phosphoprotein phosphatase activity of brain was assayed as described in the legend of Table I.

RESULTS

Marked changes in the PL-Ca-PK activity levels in different regions and fractions of rat brain were observed for the developing rat brain [25]. In the soluble and particulate fractions of grey matter (cerebral cortex plus cerebellum), the enzyme activity level progressively increases from birth to about 30 days of age, and then declines slightly in the adult (about 150 days old) (Fig. 1A). The ratio of the soluble to particulate enzyme (about 40% vs 60% of the total enzyme activity) remains rather constant throughout development. In the particulate fraction of white matter (brainstem), the level of the enzyme activity similarly increases from birth to about 30 days of age, but remains unchanged thereafter (Fig. 1B). In contrast, its level in the soluble fraction of white matter is relatively constant throughout development, resulting in a lower soluble:particulate enzyme ratio with increasing age in this brain region (Fig. 1B). These findings on rat brain are similar to the ontogeny-related changes in the enzyme activity level seen in the soluble fraction of guinea pig brain [17].

TABLE I. Phosphoprotein Phosphatase Activity in the Soluble and Particulate (Solubilized) Fraction of White and Grey Matter of Brain From Rats of Various Ages

Brain region and fraction	Age (days)	Phosphatase activity (nmol P/min/g tissue)		
		^{32}P-MBP	^{32}P-Histone	^{32}P-Protamine-SO_4
Grey matter				
Soluble	1	39.2 ± 5.2	12.2 ± 0.8	68.7 ± 5.1
	20	42.5 ± 5.7	14.8 ± 2.6	84.0 ± 5.9*
	150	52.7 ± 2.7*	12.5 ± 1.5	88.1 ± 3.7*
Particulate	1	35.7 ± 2.5	11.7 ± 1.8	61.2 ± 7.4
	20	43.7 + 2.8*	14.8 ± 0.9	74.4 ± 10.0
	150	45.3 ± 3.6*	16.8 ± 1.3*	98.1 ± 1.5*
White matter				
Soluble	1	63.3 ± 9.2	14.2 ± 2.8	75.7 ± 8.0
	20	59.5 ± 3.0	15.4 ± 0.8	87.3 ± 3.6
	150	64.3 ± 6.3	16.7 ± 0.9	84.9 ± 4.8
Particulate	1	39.6 ± 2.4	14.1 ± 0.9	64.6 ± 4.5
	20	40.9 ± 10.7	15.1 ± 1.6	74.7 ± 6.9
	150	26.1 ± 3.0*	13.1 ± 0.4	65.9 ± 5.7

Tissue fractions were prepared as in Figure 1 and aliquots (equivalent to 1.2 mg tissue) were used for the phosphatase assay. The reaction mixtures, in 0.2 ml, contained 10 μg of ^{32}P-protein (1–2 × 10^5 cpm), 10 mM $MgCl_2$, 0.25 mM $CaCl_2$, 5 μg of phosphatidylserine, and 0.5 μg of calmodulin. The reaction was carried out at 30°C for 5 min and was stopped by the addition of 2 ml of 5% trichloracetic acid/0.25% Na-tungstate, followed by the addition of 0.1 ml of 0.65% bovine serum albumin as a carrier protein. The contents of the tubes were mixed and centrifuged. An aliquot (0.5 ml) of the supernate was counted for the inorganic ^{32}P released from ^{32}P-proteins. The results obtained are shown in the table. Alternatively, the pellet was counted for the remaining ^{32}P-proteins. The results from these two methods were found to be closely comparable. The data present are mean ± SE of triple assays of brain samples (representing pools of tissue from 3–10 rats) from different age groups. *Significantly different from the day 1 group ($P < 0.05$ to $P < 0.01$). Taken from [25].

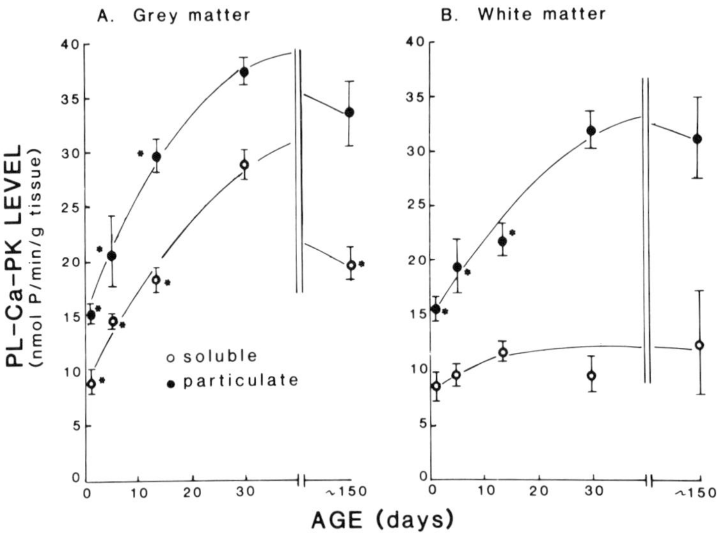

Fig. 1. Comparison of tissue levels of PL-Ca-PK in the soluble and particulate (solubilized) fractions of grey (A) and white (B) matter of brain from rats of various ages. Tissue samples (0.6–1.2 mg) were homogenized in ice-cold homogenization buffer (25 mM Tris HCl, pH 7.5/0.25 M sucrose/10 mM $MgCl_2$/50 mM mercaptoethanol). The supernate (soluble fraction) obtained by centrifuging the homogenate for 30 min at 48,000 × g was diluted 10 times with the same buffer; an aliquot (0.02 ml, equivalent to 0.4 mg tissue) was used for enzyme assay. The pellet was solubilized with the original volumes of homogenization buffer containing 0.2% Triton X-100 and 2 mM EGTA, incubated for 1 hr on ice with occasional stirring, and centrifuged at 48,000 × g for 30 min. The resultant supernate (solubilized particulate fraction) was similarly diluted 10 times, and the same aliquot (0.02 ml) was used for assay. PL-Ca-PK was assayed as described elsewhere [2,4,9,10,22]. The reaction rate was linear with respect to the amount of enzyme and time of incubation in all experiments. The activity stimulated by $CaCl_2$ (2.2- to 6.2-fold) in the presence of phosphatidylserine was presented as mean ± SE of three experiments; each experiment was carried out with brain tissue pooled from 1–5 rats. The enzyme activity stimulated by phosphatidylserine or $CaCl_2$ alone was 2 or 5% of that stimualted by a combination of both for the brain samples from all age groups. *Significantly different from the day 30 group ($P < 0.05$ to $P < 0.0001$). Taken from [25].

We reported previously that myelin basic protein (MBP) is the major endogenous substrate protein for PL-Ca-PK in the particulate fraction of brain from adult rat or guinea pig [5,6,25].. Phosphorylation of MBP in white matter (a brain region rich in myelin) was detected only after rats reach the age of 10 days, and its phosphorylation progressively becomes more pronounced as the animal matures (Fig. 2). Phosphorylation of higher M_r proteins stimulated by calmodulin/Ca^{2+}, on the other hand, remains relatively unchanged during the course of development. Absence of MBP phosphorylation in the day 1 and day 5 rat brain is not due to lack of the phosphorylating enzyme, because i-addition of exogenous MBP to the same particulate fraction of 5-day-old rats results in its marked phosphorylation, and ii-addition of exogenous PL-Ca-PK to the same brain preparation fails to demonstrate any MBP phosphorylation. The presence of PL-Ca-PK in the neonatal brain was also clearly demonstrated in Fig. 1. The development-related phosphorylation of MBP observed in the present studies (Fig. 2) is consistent with the well-recognized postnatal myelination reported by many others (eg, see [26]).

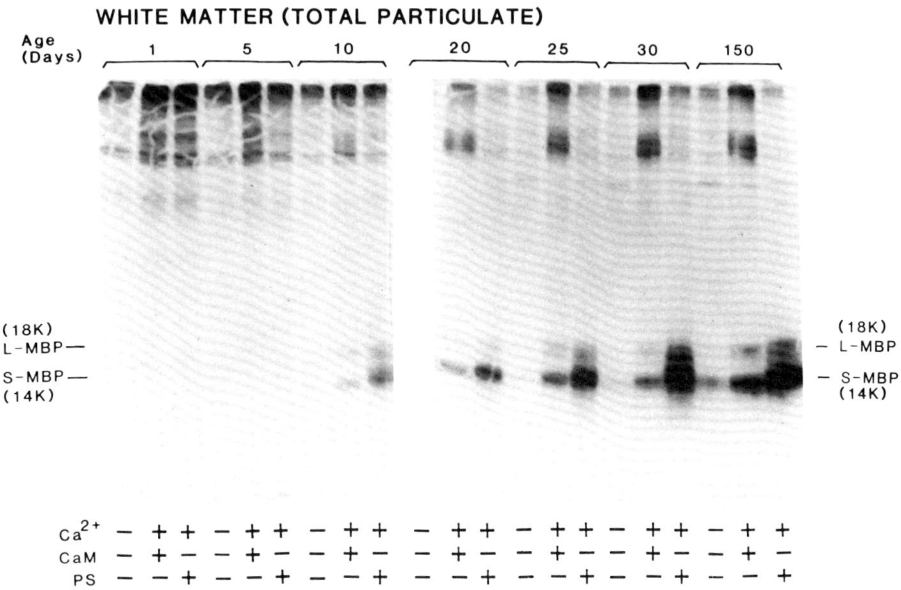

Fig. 2. Autoradiogram showing protein phosphorylation in the particulate fraction of white matter of brain from rats of various ages. The white matter was homogenized in two volumes of homogenization buffer described in Figure 1; the 48,000 × g pellet was resuspended and washed twice with the original volume of the buffer containing 10 mM EGTA (to deplete endogenous calmodulin), once with water (to cause osmotic shock), and once with the buffer. Finally, the 48,000 × g pellet was resuspended in 10 volumes of the buffer. Phosphorylation of endogenous proteins was carried out (in 0.2 ml) for 5 min, in the presence or absence of $CaCl_2$ (0.5 mM), calmodulin (CaM, 0.5 μg), and phosphatidylserine (PS, 5 μg), as indicated. EGTA (0.2 mM) was included in all cases to minimize phosphorylation caused by endogenous Ca^{2+}. The separating gel was 12% acrylamide and the amount of protein applied per lane was 10 μg. L-MBP, large MBP; S-MBP, small MBP; K, kilodalton. Taken from [25].

Next, we systematically investigated developmental changes of endogenous substrate proteins in different regions and fractions of rat brain [25]. There is only one substrate (M_r = 66K) for PL-Ca-PK in the soluble fraction of white matter, and, furthermore, its phosphorylation (perhaps reflecting the amount of protein) is the highest in the neonate (1-day old) and progressively decreases as age increases (Fig. 3A). Other substrates (M_r = 60K, 58K, and 50K), the phosphorylation of which is stimulated to a similar extent by phospholipid/Ca^{2+} or calmodulin/Ca^{2+}, on the other hand, progressively increase as a function of age. In order to detect the maximal number of substrate proteins present originally, the particulate fraction of white matter was solubilized with Triton X-100 and EGTA prior to phosphorylation. The results indicate that this solubilization step unmasked additional substrate proteins (M_r = 87K, 66K, 58K, 50K, and 46K) for PL-Ca-PK (Fig. 3B), which are otherwise undetected in the unsolubilized (total) particulate fraction ([5,6,25,27]; see also Fig. 2). It is worth noting that phosphorylation of these proteins progressively increases and reaches the highest level around day 18 and declines thereafter (Fig. 3B). Some phosphoproteins (ie, M_r = 58K, 50K, and 46K) even become nearly undetectable for the 150-day adult. MBP, migrated to the front of the gel, in contrast, increases as a function of age (Fig. 3B), consistent with the findings shown in Fig. 2. The 54K M_r

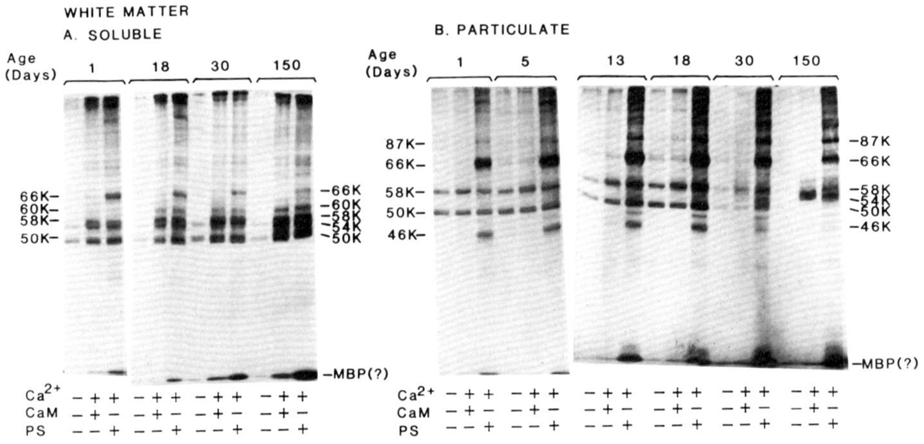

Fig. 3. Autoradiograms showing phosphorylation of proteins from the soluble (A) and particulate (solubilized) (B) fractions of white matter of brain from rats of various ages. Tissue fractions were prepared as described in Figure 1. The phosphorylation conditions were the same as in Figure 2, except that the incubation time was 2 min, the separating gel was 8% acrylamide (to improve resolution of higher M_r proteins), and the particulate fraction was subjected to prior solubilization with Triton X-100 and EGTA as described in Figure 1. Taken from [25].

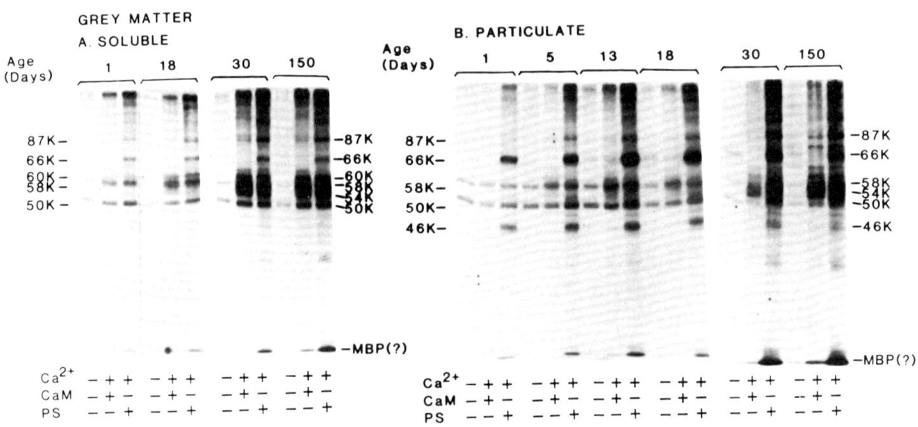

Fig. 4. Autoradiograms showing phosphorylation of proteins from the soluble (A) and particulate (solubilized) (B) fractions of grey matter of brain from rats of various ages. Experimental conditions were the same as described in Figure 3. Taken from [25].

protein, a common substrate for the calmodulin/Ca^{2+}-dependent species of protein kinase and PL-Ca-PK, is clearly detectable only for the 150-day-old rat (Fig. 3B).

The phosphorylation patterns of endogenous proteins from grey matter (Fig. 4) are similar to those from white matter shown in Fig. 3. Certain differences, however, exist: in the soluble fraction of grey matter, the 66K M_r protein increases as a function of age, and the 87K M_r protein appears to be specific to this brain region (Fig. 4A);

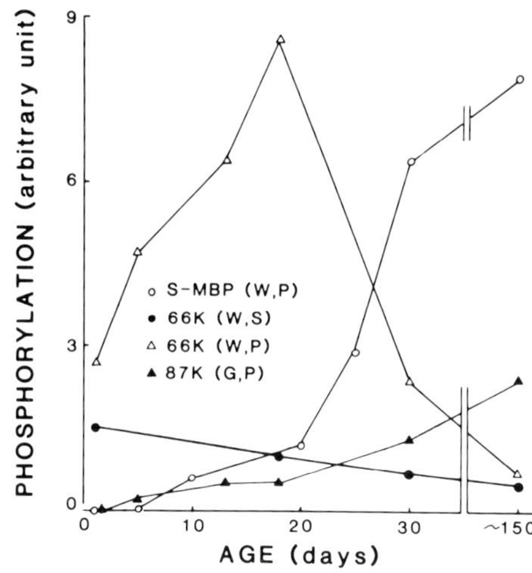

Fig. 5. Phosphorylation of representative endogenous proteins from white and grey matter of brain from rats of various ages. Densitometric tracings of the phosphatidylserine/Ca^{2+}-dependent phosphorylation of small MBP (S-MBP) was taken from Figure 2; the 66K M_r protein in soluble fraction of white matter was from Figure 3A; the 66K M_r protein in particulate fraction of white matter was from Figure 3B; and the 87K M_r protein in particulate fraction of grey matter was from Figure 4B. W, white matter; G, grey matter; P, particulate fraction; S, soluble fraction. Taken from [25].

in the particulate (solubilized) fraction, the 58K and 50K M_r proteins remain maximally phosphorylated even at the age of 150 days (Fig. 4B).

Developmental changes in certain representative substrates in brain for PL-Ca-PK shown in Figures 2–4 are semi-quantitated by densitometric scanning of the autoradiograms on the original X-ray film (Fig. 5). Phosphorylation of small MBP (14K, in the particulate fraction of white matter) and the 87K protein (particulate, grey matter) increases, whereas the 66K M_r protein (soluble, white matter) decreases as a function of age. In comparison, apparently the same 66K M_r protein (particulate, white matter) shows a biphasic change, reaching a peak level at day 18. The data shown in Figure 5 also indicate a marked change in the soluble:particulate ratio of the 66K M_r protein in white matter during development. Thus, the ratio was 0.6 at day 1, 0.1 at day 18, and returned to 0.7 at day 150. The significance of the apparent "translocation" of the protein, however, is unclear at present.

We also examined the developmental change of an endogenous substrate protein (M_r = 19K) in brain for the S-100-modulated phosphorylation system (Fig. 6). The data indicate that the substrate for protein kinase X is present at a far higher amount in the adult (150-day old) than in the neonate (2-day old). As seen for the adult brain, the substrate protein in the neonatal brain becomes detectable only after the step of S-100 affinity chromatography [24]. It seems worth noting that the 40K M_r phosphoprotein is predominant in the neonate, whereas the 22K M_r protein is predominant in the adult (Fig. 6). Although the species of protein kinases that phosphorylate these proteins are unclear at present, it is conceivable that the 40K M_r protein may be a "neonatal protein" more critically involved in brain development, whereas the 22K

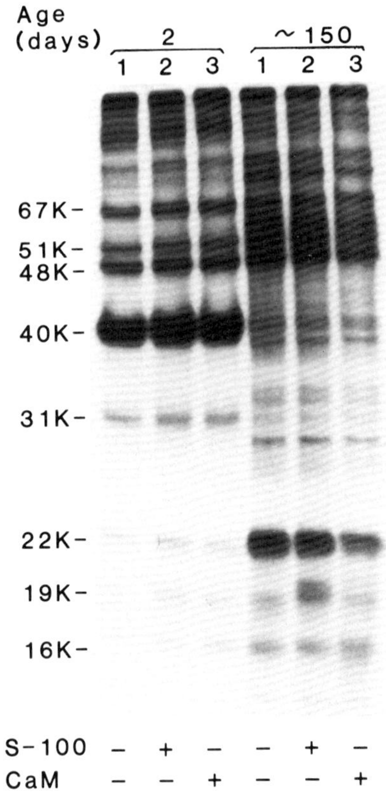

Fig. 6. Autoradiogram showing the developmental change of an endogenous substrate protein (M_r = 19K) for protein kinase X in an S-100 affinity fraction of the total solubilized fraction of brain from rats of 2 days (neonate) and 150 days (adult) of age. Taken from [24].

M_r protein, as the 19K M_r substrate for protein kinase X, may be an "adult protein" more crucial for brain maturation.

A developmental study of protein kinase X in brain was also conducted in order to correlate the change in the enzyme activity level with that in the substrate level shown in Figure 6. It was noted that the enzyme activity increases in both the soluble and particulate (solubilized) fractions of brain as a function of age, reaching a peak at day 30 and remaining unchanged thereafter (Fig. 7). The ratio of the soluble:particulate enzyme activity remains rather constant throughout development and maturation, with the exception of the earliest neonatal stage (day 2) studied. A low degree of phosphorylation of the 19K M_r protein seen above in the neonatal brain (Fig. 6) is probably due to a deficiency in the substrate but not the phosphorylating enzyme, because the addition of exogenous protein kinase X to the same neonatal preparation did not result in a higher phosphorylation of the protein.

In order to assess the contribution of dephosphorylation enzymes to the state of endogenous protein phosphorylation, developmental changes in the phosphoprotein phosphatase activity were investigated [25]. Because the species of phosphatases involved are unclear, three phosphoproteins (ie, ^{32}P-MBP, ^{32}P-histone, and ^{32}P- pro-

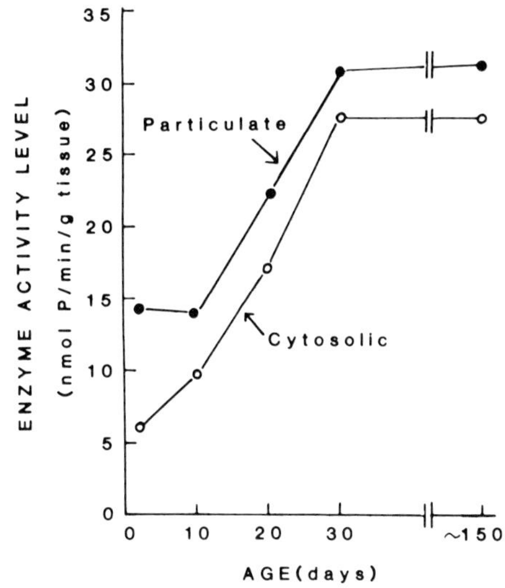

Fig. 7. Comparison of tissue levels of protein kinase X in the soluble (cytosolic) and particulate (solubilized) fractions of brains from rats of various ages. Preparations of the fractions and the assay from the enzyme activity were as described [23]. The net increase in protamine phosphorylation promoted by S-100 (about 3.2- to 4.1-fold) was used for calculating the enzyme activity. The data presented are the averages of duplicate incubations, with assay error being less than ± 5%. The brain samples used were pooled from 3–6 rats of individual age groups. Taken from [24].

tamine-SO_4) were used as substrates, and the enzyme activity was assayed in the presence of Ca^{2+}, calmodulin, and phosphatidylserine, conditions resembling those used for endogenous protein phosphorylation. It was found that the phosphatase activity remains relatively unchanged in brain tissue throughout development (Table I). Only a small increase (up to 60%) in the soluble and particulate fractions of grey matter and a small decrease (52%) in the particulate fraction of white matter from adult rat were noted. The results were markedly different from those seen for PL-Ca-PK (Fig. 1) and protein kinase X (Fig. 7) activities and phosphorylation of endogenous proteins (Figs. 2–4,6). It appears, therefore, that the phosphorylation reaction, but not the dephosphorylation reaction, may be more crucial in determining the overall extent of protein phosphorylation and thus the function of certain proteins. It was reported recently that the calmodulin/Ca^{2+}-stimulated species of phosphatase, as determined by radioimmunoassay, increases 2–4 times postnatally in rat brain [28]. Because the activity of this enzyme relative to the total phosphatase activity in brain is unknown, its contribution to the maintenance of the overall phosphorylation state of cellular proteins remains a matter of speculation at present.

DISCUSSION

The findings presented here clearly show certain major developmental changes in protein phosphorylation systems of rat brain. The results obtained with PL-Ca-PK (Fig. 1) agree with our previous findings on the enzyme levels in guinea pig brain [17]. The ontogeny-related changes in PL-Ca-PK in brain are different from those previously seen for the cyclic AMP- and cyclic GMP-dependent protein kinases [29], supporting the notion of distinct functional and regulatory roles for these enzymes. It is of interest that cytosolic PL-Ca-PK levels increase markedly as F9 teratocarcinoma cells are stimulated to differentiate to a parietal cell type [30], a situation perhaps analogous to developing brain tissue. The ontogenetic changes in PL-Ca-PK levels seen in the guinea pig are similar to those of rat brain, except for the shift in time scale. In the guinea pig, PL-Ca-PK activity increases most rapidly in the 20 days before birth and reaches a maximal level at birth [17]. In the rat, however, the period of rapid increase in PL-Ca-PK activity occurs postnatally, and reaches a maximum at about 30 days of age (Fig. 1). It is of interest that the peak rate of brain mass growth in guinea pig occurs at 18 days before birth, while the peak rate of brain growth in rat is about 8 days after birth. Thus, for both species, the increase in PL-Ca-PK levels coincides with those of brain maturation (synaptogenesis and myelinogenesis), and these levels reach maximum about 20 days after the peak in rate of brain growth. The level of protein kinase X also increases with brain development (Fig. 7), again suggesting a role for this new phosphorylation system in maturation of brain.

Recent findings indicate that PL-Ca-PK appears to be a phorbol ester receptor protein [31–34]. It is of interest that the highest level of phorbol ester receptors have also been found in the matured brain of the developing chicken embryo [35], in agreement with the findings made with PL-Ca-PK in developing rat brain (present studies) or guinea pig brain [17].

The Ca^{2+}-dependent phosphorylation of endogenous protein in the soluble fraction of brain can be clearly demonstrated by using directly the preparation for phosphorylation (Figs. 3A and 4A). The phosphorylation patterns, however, are slightly different from those obtained in previous studies [6], in which the fraction was treated to deplete endogenous calmodulin. No substrate protein specific for the calmodulin/Ca^{2+}-dependent enzyme, however, was detected in both instances. It is further noted that no substrate specific for the calmodulin/Ca^{2+}-dependent enzyme was detected in the soluble fraction of brain from rats of various ages, suggesting that PL-Ca-PK is a predominant protein phosphorylation system throughout the development process of the brain.

The findings showing distinct changes in various endogenous substrates for protein phosphorylation systems regulated by phospholipid/Ca^{2+}, calmodulin/Ca^{2+} and S-100 in brain during development are intriguing. These substrate proteins can be classified into two major categories: (1) Proteins that increase steadily and are present at the highest levels in the adult. MBP and the 87K, 58K, 54K, and 50K M_r substrate proteins for PL-Ca-PK and the 19K M_r substrate protein for protein kinase X were the examples, and they may be referred to as "adult proteins" involved in mature brain function. (2) Proteins that exhibit a biphasic developmental change, with a peak level closely following the peak growth rate of brain mass. These proteins, exemplified by the 66K and 46K M_r proteins phosphorylated by PL-Ca-PK, may be termed "neonatal proteins" intimately involved in the development of brain. It is well recognized that sequential replacement of discrete cellular components, including proteins, is a normal, general, and important event of ontogeny of muscle [36]. The evidence obtained from these studies strongly suggests that this replacement process also occurs in developing brain, and that protein phosphorylation systems are a part of this replacement process. Investigations into the detailed subcellular and ultrastructural localization of protein kinases and their substrate proteins would undoubtedly yield insights into the functional significance of protein phosphorylation systems in brain ontogeny.

ACKNOWLEDGMENTS

These studies were supported by USPHS Grants HL-15696, NS-17608, CA-36777, and T32-GM-07594.

REFERENCES

1. Takai Y, Kishimoto A, Iwasa Y, Kawahara Y, Mori T, Nishizuka, Y: J Biol Chem 254:3692, 1979.
2. Wise BC, Raynor RL, Kuo JF: J Biol Chem 257:8481, 1982.
3. Kikkawa U, Takai Y, Minakuchi R, Inohara S, Nishizuka Y: J Biol Chem 257:13341, 1982.
4. Schatzman RC, Raynor RL, Fritz RB, Kuo JF: Biochem J 209:435, 1983.
5. Wrenn RW, Katoh N, Schatzman RC, Kuo JF: Life Sci 29:725, 1981.
6. Wrenn RW, Katoh N, Wise BC, Kuo JF: J Biol Chem 255:12042, 1980.
7. Katoh N, Wrenn RW, Wise BC, Shoji M, Kuo JF; Proc Natl Acad Sci USA 78:4813, 1981.
8. Sano K, Takai UY, Yamanishi J, Nishizuka Y: J Biol Chem 258:2010, 1983.
9. Helfman DM, Barnes KC, Kinkade JM Jr, Vogler WR, Shoji M, Kuo JF: Cancer Res 43:2955, 1983.
10. Wise BC, Glass DB, Chou C-HJ, Raynor RL, Katoh N, Schatzman RC, Turner RS, Kibler RF, Kuo JF: J Biol Chem 257:8489, 1982.
11. Turner RS, Chou C-HJ, Kibler RF, Kuo JF: J Neurochem 39:1397, 1982.
12. Katoh N, Wise BC, Kuo JF: Biochem J 209:189, 1982.
13. Mazzei GJ, Kuo JF: Biochem J 218:361, 1984.
14. Kuo JF, Schatzman RC, Turner RS, Mazzei GJ: Mol Cell Endocrinol 35:65, 1984.
15. Schatzman RC, Turner RS, Kuo JF: In Cheung WY (ed): "Calcium and Cell Function." New York: Academic Press, Vol. 5, 1984, p. 33.
16. Turner Rs, Mazzei GJ, Kuo JF: In Kuo JF (ed): "Phospholipids and Cellular Regulation." Boca Raton: CRC Press, 1984, Vol. 2, (in press).
17. Wise BC, Andersson RGG, Mackerlova L, Raynor RL, Solomonsson I, Kuo JF: Biochem Biophys Res Commun 99:407, 1981.
18. Moore BW: Biochem Biophys Res Comm 19:739, 1965.
19. Moore BS: Intl Rev Neurobiol 15:215, 1972.
20. Bock E: J Neurochem 30:7, 1978.
21. Baudier J, Briving C, Deinum J, Haglid K, Sarskog L, Wallin M: FEBS Lett 147:165, 1982.
22. Kuo JF, Andersson RGG, Wise BC, Mackerlova L, Solomonsson I, Bracketrt NL, Katoh N, Shoji M, Wrenn RW: Proc Natl Acad Sci USA 77:7039, 1980.
23. Qi DF, Turner RS, Kuo JF: J Neurochem 42:458, 1984.
24. Qi DF, Kuo JF: J Neurochem 43:256, 1984.
25. Turner RS, Raynor RL, Mazzei GJ, Girard PR, Kuo JF: Proc Natl Acad Sci USA 81:3143, 1984.
26. Benjamins JA, McKhann GM: In Siegel GJ, Albars RW, Agranoff BW, Katzman R (eds): "Basic Neurochemistry" Boston: Little, Brown and Co., 1981, p 445.
27. Mazzei GJ, Qi DF, Schatzman RC, Raynor RL, Turner RS, Kuo JF: Life Sci 33:119, 1983.
28. Tallant EA, Cheung WY: Biochemistry 22:3630, 1983.
29. Kuo JF: Proc Natl Acad Sci USA 73:2256, 1975.
30. Kraft AS, Anderson WB: J Biol Chem 258:9178, 1983.
31. Castagna M, Takai Y, Kaibuchi K, Sano K, Kikkawa U, Nishizuka Y: J Biol Chem 257:7847, 1982.
32. Niedel JE, Kuhn LJ, Vandenbark GR: Proc Natl Acad Sci USA 80:36, 1983.
33. Ashendel CL, Staller JN, Boutwell RK: Biochem Biophys Res Comm 111:340, 1983.
34. Kikkawa U, Takai Y, Tanaka Y, Miyake R, Nishizuka Y: J Biol Chem 258:11442, 1983.
35. Nagle DS, Jaken S, Castagna M, Blumberg PM: Cancer Res 41:89, 1981.
36. Caplan AI, Fiszman MY, Eppenberger HM: Science 221:921, 1983.

Neurobiology: Molecular Biological Approaches to Understanding Neuronal Function and Development 113–120 (1985)

Non-Random Opening of Voltage-Dependent Potassium Channels in Cultured Hippocampal Neurons

Michael A. Rogawski

Laboratory of Neurophysiology, National Institute of Neurological and Communicative Disorders and Stroke, National Institutes of Health, Bethesda, MD 20205

Voltage-dependent potassium channels were recorded from the somatic membranes of cultured rat hippocampal neurons using the gigaseal patch technique in the cell-attached configuration. Depolarizing voltage steps activated outward single-channel currents that occurred in bursts with a mean lifetime of 10 msec (for openings $<$ 30 msec) and a mean single channel conductance of 20 pS. The frequency of channel opening increased markedly with depolarization in the range 15 to 50 mV. Reversal potential measurements demonstrated that the channels carry predominantly potassium ions. Statistical analysis of the occurrence of bursts during repeated depolarizing voltage steps indicated that the potassium channels open in a non-random fashion with time. The frequency distribution of the number of bursts per step deviated markedly from that predicted by a Poisson distribution due to the presence of a greater than expected number of steps with no channel openings. In addition, the runs test demonstrated that, for some channels, periods of channel opening are clustered non-randomly between stretches of records with no openings. Non-random opening as observed in these experiments is a novel property of potassium channel gating that cannot be detected in conventional recordings and indicates the slow transition of the channel into and out of a voltage-independent inactivated state.

Key words: **cultured hippocampal neurons, single-channel recording, patch clamp, potassium channel**

Direct observation of the unitary currents flowing through single ion channels with the patch-clamp technique has revealed complexities of the channel gating mechanism that could not be appreciated from the study of macroscopic currents under voltage clamp or by the analysis of current fluctuations ("noise"). One such novel property is the tendency of some channels to open in a non-random fashion with time. This was initially observed in studies on extrasynaptic acetylcholine receptor channels from frog muscle [1] and in planar membranes seeded with acetylcholine receptors isolated from *Torpedo* [2] where, in the presence of high agonist concentrations, channel opening occurred in "clusters" of bursts often separated by several seconds in the closed state. This clustering is believed to represent the single-channel correlate of the decline in muscle endplate current ("desensitization") occur-

Received May 4, 1984.

ring in the continued presence of acetylcholine. More recently, studies of glutamate-activated channels in locust muscle have demonstrated that the frequency of opening in successive time intervals also deviates significantly from that predicted for a random process [3]. It has been suggested that the non-random activity of these chemically gated channels resides either in temporal variations in ligand binding, in the coupling of the receptor to the ion channel, or in the channel itself [3].

Voltage-activated sodium channels in clonal pituitary cells have also been observed to show non-random fluctuations in opening frequency over a slow time scale. When stimulated by repetitive depolarizing pulses, these channels seem to cycle into and out of a non-activatable state. This is observed as a non-random clustering of records with and without channel openings [4].

In this communication I demonstrate that voltage-dependent potassium channels in the somata of cultured hippocampal neurons display spontaneous fluctuations in opening frequency. These channels, which in some respects resemble the potassium channels of squid axon [5], are rather sparsely distributed over the somatic membrane of hippocampal neurons, so that it is common to obtain one or two channels per membrane patch. This situation is convenient in that it allows the stochastic properties of individual ion channels to be studied in a simple fashion. Patch clamp recordings clearly indicate that the opening frequency of somatic voltage-dependent potassium channels vary in a non-random fashion with time. Therefore, slow transitions to a non-activatable state are a property of some voltage as well as ligand-gated channels. For voltage-activated channels, this appears to be an intrinsic property of the channel molecule itself, rather than a characteristic of the transmitter-receptor interaction or the coupling of the receptor to the ion channel.

MATERIALS AND METHODS

Neurons dissociated from the hippocampii of 19-day-old rat embryos were grown in culture according to methods developed by Segal [6]. Cells maintained in culture for less than 2 wks were used for patch-clamp recording. Prior to each experiment the cells were washed with recording medium containing (in mM): 137 NaCl, 5.5 KCl, 0.40 KH_2PO_4, 0.34 $NaHPO_4$, 2–4 $CaCl_2$ 2–4 $MgCl_2$, 12.5 HEPES, and sufficient glucose to bring the osmolality to 310–320 mOsm/l (pH 7.4). Tetrodotoxin (1–2 μM) and $CdCl_2$ (0.2 mM) were routinely added to the recording medium to block voltage-dependent sodium and calcium channels, respectively.

Gigaseal recordings were carried out essentially according to methods described by Hamill et al [7] using a Dagan 8900 patch clamp amplifier (Dagan Corp, Minneapolis, MN). Flint glass, silicone elastomer (Sylgard 184, Dow Corning, Midland, MI) coated pipettes were used. When fire polished and filled with bath solution, the impedance of the pipettes was typically 2.5 to 5 MΩ. Experiments described in this paper were all carried out in the cell-attached configuration at room temperature (21 to 24° C).

Depolarizing voltage commands 100 to 300 msec in duration were applied to the patch pipette at a rate of 0.2 to 0.3 Hz. The patch was depolarized by stepping the interior of the pipette to a negative potential with respect to the bath. For convenience, the magnitude of the step potential is reported as a positive number in the text and figures.

The current records were filtered at 1 kHz, digitized at 100 to 200 μsec/point, and stored on the Winchester disk of a minicomputer system for later analysis.

Individual channel records were manually analyzed to obtain current jump transitions and amplitudes on a video display monitor using a cursor program. Capacitative transients occurring at the beginning and end of the voltage steps were eliminated by subtracting the average of a series of records containing no channel openings. The transient was typically 3 msec or less in duration.

Statistical analysis was carried out on channel frequency data derived from an ensemble of identical voltage steps. The expected Poisson distribution was calculated from the formula shown in Figure 2 using the experimentally observed mean number of bursts per step. Comparisons between the observed and expected distribution were made with the chi square test. The run test is as described by Siegel [8]. For small samples (patch 2, Table I), a table of critical values of r, the number of runs, was used to determine significance at the 0.05 level. For large samples (number of records of one kind > 20), the standardized random variable z was estimated according to the formula $z = (r - \mu_r) / \sigma_r$ where μ_r and σ_r are the calculated mean and standard deviation of the sampling distribution of r, which approximates the normal distribution (see Siegel [8]). The significance of the computed value of z was determined from the one-tailed probabilities associated with values as extreme as z in the normal distribution.

RESULTS

With the patch pipette at bath potential, so that the transmembrane potential equaled the cell resting potential, unitary current jumps were rarely observed. However, upon step depolarization of the patch by 15 to 60 mV, outward unitary currents were invariably recorded (Fig. 1A). It was immediately obvious that the frequency of these events fluctuated in time, so that during certain periods of the experiment each record in a series of voltage steps contained multiple channel openings, whereas at other times no channel opening occurred. On average, for many steps, the probability of channel opening increased strongly with voltage, but even for large depolarizations the channels were open only a small fraction of the time. Although the record shown in Figure $1A_1$ to 50 mV shows the channel to be almost continuously in the open state, many companion steps in the same series had no channel openings, so that the ensemble probability at this potential was 0.023. At a given potential, channel

TABLE I. Statistical Analysis of Burst Frequency During Repetitive Voltage Steps

Patch	Step potential (mV)	Step duration (msec)	Number of bursts	Deviation from Poisson ($P <$)	Deviation from randomness by runs test[a] z	P
1	12	100	73	0.02		
2	20	300	36	0.005	−0.97	NS
3	20	300	69	0.05	−2.2	0.013
4	24	200	138	0.001	−7.5	<<0.001
5	40	300	36	0.005	−2.2	0.013
6	46	100	47	0.005	−4.5	<<0.001
7	50	100	64	0.001	−1.0	NS

[a]The standardized random variable z was calculated as described in Materials and Methods. NS, not significant.

amplitudes were normally distributed (Fig. 1B), although in some cases there was a slight skewing of the histogram toward lower amplitudes, as is shown in Figure 1B. There was no evidence for subconductance states.

Channel amplitudes varied linearly with potential (Fig. 1C). For four typical channels in separate membrane patches the single channel slope conductance was 20 ± 2 pS (mean ± SD). The extrapolated reversal potential was 21 ± 5 mV below rest. Assuming a cell resting potential of −50 to −60 mV, the true reversal potential is therefore −70 to −80 mV, identical to that obtained for the reversal potential of macroscopic potassium currents in these cells (see Segal and Barker [9]). Experiments with excised patches, where the solution in contact with the internal membrane face can be varied, confirm that the reversal potential is well predicted by the Nernst relationship for potassium, indicating that the channels carry predominantly potassium ions [manuscript submitted].

Channel gating was occasionally interrupted by brief closures (Fig. $1A_2$), so that open events could naturally be grouped into "bursts." Bursts were defined as successive channel openings separated by closed periods of no greater than 4 msec. There were typically 0.5 to 0.6 resolvable closings (current record descending to baseline) per burst, and slightly more than one-half of the bursts had no intermediate closing events that were detected. Bursts lifetimes less than 30 msec in duration were exponentially distributed, with a time constant of 10 ± 5 msec. A disproportionately large number of longer bursts were observed, indicating that there may be a second kinetic component to the burst lifetime distribution.

During the repetitive application of depolarizing commands to a patch of membrane, the frequency of channel opening fluctuated with time. This is illustrated in Figure 2A, where the number of bursts per depolarizing step (200 msec) is shown for 168 successive steps. Most frequently, no bursts occur during the depolarizing step, but when a burst does occur there may be up to 10 per step. It is unlikely that fluctuations in the seal between the pipette and membrane account for the variability in the number of channel openings per step, as the small capacitative transients occurring at the beginning and end of the step and the leakage current remain stable throughout the recording. Cumulative inactivation as defined by Aldrich et al [10]—that is, inactivation continuing during repolarization—cannot account for the fluctuation in channel opening. This is because periods of high frequency opening can begin late in a series of steps following multiple records with no openings. Conditioning hyperpolarization of the patch by up to 30 mV did not reduce the tendency of channels to open in a non-random fashion with time.

Two statistical methods were used to examine whether the occurrence of bursts in an ensemble of depolarizing steps could be described by a random process. In the first method, a comparison was made between the observed frequency distribution of the number of bursts per step and that predicted under the assumptions of a Poisson-process. The Poisson distribution should apply if the probability of burst occurrence per step is small and constant in time. The observed and predicted distributions for the experiment of Figure 2A are shown in Figure 2B. Note that there is a disproportionately large number of steps with zero openings and a correspondingly larger than expected number of steps with many openings. This pattern was repeated in all patches examined. Table I summarizes the results from experiments on seven me-patches, including the one shown in Figure 2. In each of these experiments, double openings did not occur, suggesting that only a single channel was present in each

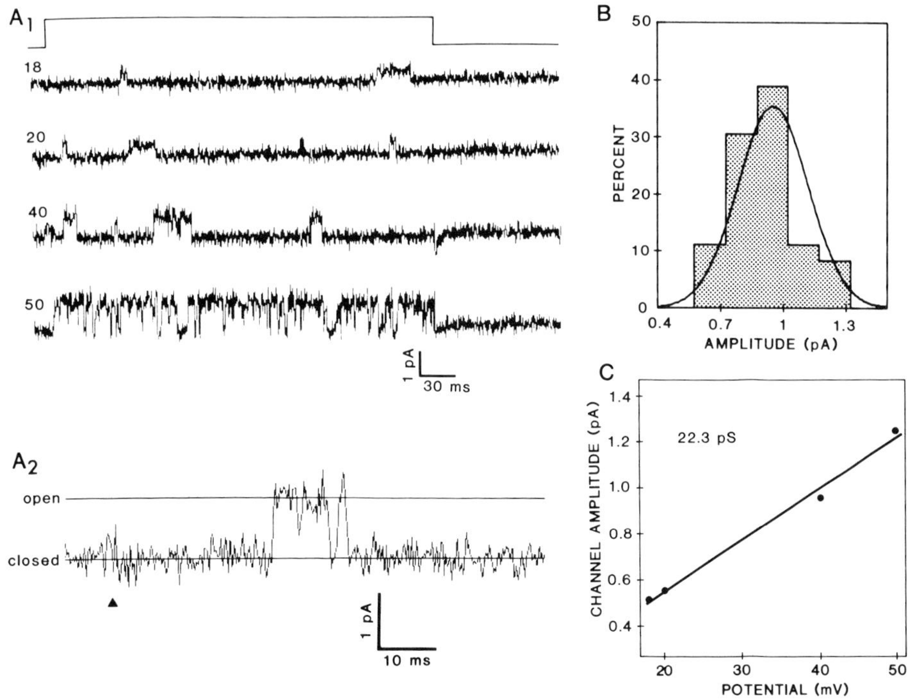

Fig. 1. Single-channel currents recorded from the somatic membrane of a cultured hippocampal neuron. A_1) Unitary currents obtained at various potentials during depolarizing step commands. The duration of the step is indicated at the top of the figure. Capacitative transients have been subtracted as described in Materials and Methods. A_2) Example of a burst from the same membrane patch showing one (and possibly a second) brief closure. The small triangle indicates onset of the depolarizing voltage step of 40 mV. B) Histogram showing the distribution of channel amplitudes for a depolarization of 40 mV. The curve is the appropriately scaled normal distribution with mean (0.95 pA)and standard deviation (0.17pA) derived from the experimental data. C) Plot of single channel amplitude versus step potential. The single-channel conductance is derived from the slope of the best fit line.

patch. However, the openings were sufficiently infrequent that the statistical tests devised by Colquhoun and Hawkes [11] could not be used to confirm this point. Nevertheless, if more than one channel exists in a patch this would tend to decrease the non-randomness, so that the analysis, if anything, would underestimate the degree of deviation from random behavior. For some patches, channel opening ceased completely for a period of time, and the experiment was terminated. To provide the most stringest statistical test, "trailing" zero records prior to termination of an experiment were not included in the statistical analysis.

Chi-square comparison of the expected and predicted distributions showed a significant deviation from that predicted by Poisson statistics for all patches examined. This indicates that channel opening within an individual step is highly correlated, but does not address the issue of whether there is a relationship between channel opening probability in temporally close steps. To determine if there is a tendency for steps with one or more channel openings to cluster in time, the non-parametric runs test was applied. The entire series of steps in sequence (except those trailing) were scored

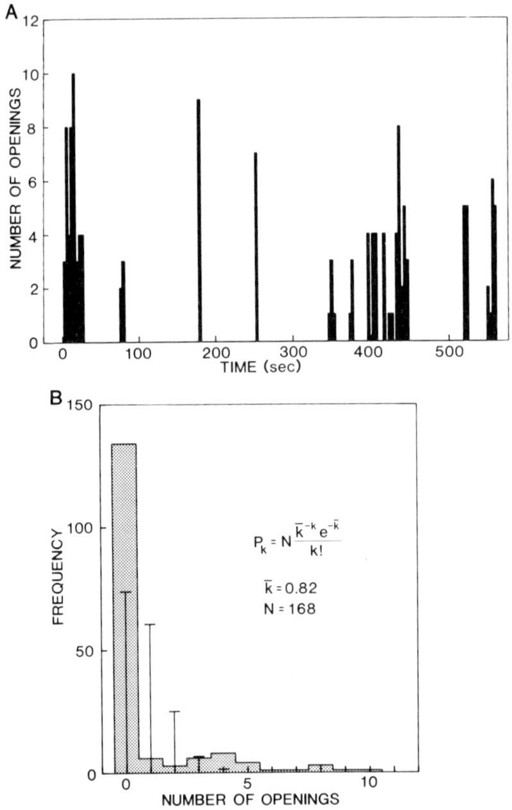

Fig. 2. Non-random opening of potassium channels during repetitive voltage steps. A) Histogram showing number of bursts per depolarizing step (24 mV) in a series of 168 successive steps applied at 0.5 Hz. The gigaseal was achieved several seconds before zero time. B) Comparison of the observed frequency distribution (shaded bars) with that predicted by a Poisson distribution ("T"-shaped lines). The data is from the patch shown in A. The predicted distribution was calculated according to the equation shown, where N is the number of steps, $\bar{k}$ is the mean number of openings per step, and k is an integer number of openings. According to the chi square test with two degrees of freedom, the distributions differ with $P < 0.001$.

either as empty or as containing one or more bursts. For many membrane patches, the runs test indicated that there was a non-random clustering of steps, but not all experiments showed this property (see Table I). In all experiments (even those not reaching statistical significance), z was less than zero, indicating that the observed number of runs is less than the expected value; ie, there is a clustering of records with at least one channel opening. If trailing empty records are included in the analysis, all patches show a significant deviation from randomness by the runs test.

DISCUSSION

The somatic membrane of hippocampal neurons in tissue culture contain voltage-sensitive potassium channels, the properties of which are similar in some respects to those of the delayed rectifier channels of squid axon [5] or lobster nerve [12]. Preliminary studies indicate that these channels show a slow increase in opening probability after an instantaneous change in membrane potential, that opening is independent of calcium entry, and that the channels are sensitive to tetraethylammonium.

The experiments described here demonstrate that these channels open in a non-random fashion during repeated voltage steps. This suggests that the channels can enter a long-lived inactivated state, which persists even during repolarization. Although the channels do spontaneously exit from the inactivated state, this often requires several seconds or more. The tendency to leave the long-lived inactivated state seems independent of membrane potential and is therefore distinct from well-recognized inactivation processes. Nevertheless, the relationship between conventional time-dependent inactivation and the fluctuations in opening frequency described here is, at present, unknown.

On the basis of these single channel recordings, it can only be speculated that the long-lived inactivated state occurs normally in nerve cell membranes. It is possible that perturbation of the channel's membrane environment produced by the patch pipette leads to the occurrence of the inactivated state. Alternatively, the channel may move laterally in the membrane, so that at times during the experiment current flow is blocked by the glass of the patch pipette. However, if longlasting inactivation is an intrinsic property of neuronal potassium channels, this is a characteristic that is undetectable by conventional neurophysiological methods. It has been suggested that the non-random opening of a glutamate-activated cation channels in locust muscle results from a change in the affinity of the glutamate binding site with time [3]. This does not apply in the present situation of agonist-independent gating, and it must be assumed that if non-random opening reflects a physiologically relevant property of potassium channels this is intrinsic to the channel or its gating mechanism. (Preliminary experiments with isolated excised patches suggest that the phenomenon is not related to underlying cell activity.) Whether this inactivation process can be physiologically modulated to alter the magnitude of the macroscopic potassium conductance remains to be determined.

In summary, these experiments demonstrate that at least three kinetic processes control the gating of voltage-dependent potassium channels in hippocampal neurons. Burst opening corresponds most closely to the usual conception of channel gating based upon the naive assumptions held prior to the advent of single-channel recording. The ability to record single-channel currents has revealed the existence of two additional kinetic processes in the gating of these channels. The first is the occasional rapid (few millisecond) transition of the channel into and out of the closed state during a burst, referred to as "flickering." The second process, as described here, is the prolonged (several seconds) transition of the channel into an inactivated state. These processes, in concert, determine the magnitude and kinetics of the macroscopic potassium conductance.

ACKNOWLEDGMENTS

I am indebted to G.D. Lange for suggestions regarding the statistical anaylsis and assistance with the computer programming, to J.L. Barker for support and encouragement, and to J. Mazzetta and V. Smallwood for preparing the tissue cultures. The author was a Medical Staff Fellow in the Pharmacological Sciences Program, National Institute of General Medical Sciences.

REFERENCES

1. Sakmann B, Patlak J, Neher E: Nature 286:71–73, 1980.
2. Schindler H, Quast U: Proc Natl Acad Sci USA 77:3052–3056, 1980.
3. Gration KAF, Lambert JJ, Ramsey R, Usherwood PNR: Nature 291:423–425, 1981.
4. Horn R, Vandenberg CA, Lange K: Biophys J 45:323–335, 1984.
5. Conti F, Neher E: Nature 285:140–143, 1980.
6. Segal M: J Neurophysiol 50:1249–1264, 1983.
7. Hamill OP, Marty A, Neher E, Sakmann B, Sigworth FJ: Pflügers Arch 391:85–100, 1981.
8. Siegel S: "Nonparametric Statistics for the Behavioral Sciences." New York: McGraw-Hill, 1956.
9. Segal M, Barker JL: J Neurophysiol 51:1409–1433, 1984.
10. Aldrich RW, Getting PA, Thompson SH: J Physiol 291:507–530,1979.
11. Colquohoun D, Hawkes AG: In Sakmann B, Neher E (eds): "Single-Channel Recording." New York: Plenum Press, 1983, pp 135–175.
12. Coronado R, Latorre R, Mautner HG: Biophysical J 45:289–299, 1984.

Neurobiology: Molecular Biological Approaches to
Understanding Neuronal Function and Development 121–125 (1985)

Viruses and the Central Nervous System: Workshop Report

Ashley T. Haase

Infectious Disease Section, Veterans Administration Medical Center, San Francisco, California 94121

Most of what we know about the interactions of viruses and cells in the central nervous system (CNS) is based on observations of neurological diseases caused by viruses. I will develop this theme first, and then describe how investigation of virus gene expression in chronic neurological infections has revealed new subtleties in the relationships of viruses and cells that deepen our appreciation of the ecology of virus infections. In the course of this description, I will emphasize the utility of hybridization methods that should be applicable to a number of problems in neurobiology, particularly three new techniques for: 1) simultaneous detection of genes and proteins in the same cell; 2) detection of two genes by double label in situ hybridization and color microscopic radioautography; and 3) detection of virus genes in whole animals or organs by a combined macroscopic-microscopic hybridization method, in which the fine cellular resolution of in situ hybridization is preserved.

VIRUSES AND NEUROPATHOLOGY

When viruses replicate in cells in culture they may kill or transform the cell, alter the metabolism of the cell, or produce no detectible change. The corresponding effects of viruses in the CNS are as follows:

Direct Destructive Effects

Viruses like polio kill neurons and produce an acute paralytic disease. Chronic neurological disease also can result from direct cytocidal interactions of viruses and cells in the CNS. In progressive multifocal leucoencephalopathy virus replication in oligodendrocytes causes a demyelinating disorder that develops over months.

Indirect Destructive Effects: Immunopathology

In the majority of cases, however, the damage viruses induce are indirectly mediated and involve the host immune system. The target antigens may be either viral or cellular.

Received May 10, 1984.

Viral Antigens; Identification of Cellular Targets by Simultaneous Detection of Genes and Proteins in the Same Cell. In visna, a slowly progressive paralytic disease of sheep, the immune and inflammatory response to cells bearing viral antigens leads to the destruction of infected cells and "innocent bystanders" in the inflammatory focus. Demyelination is the consequence of destruction of neural and glial elements, oligodendrocytes in particular. To unequivocally establish that the latter cell type is one of the cellular targets for the immune response, we have developed and utilized a method that detects nucleic acids and proteins in the same cell. In this case an antiserum to sheep oligodendrocytes was used first with peroxidase and diaminobenzidine (DAB) to deposit an insoluble polymer of DAB on the surface of oligodendrocytes. This polymer will withstand subsequent in situ hybridization procedures with a radiolabelled probe specific for visna virus. In the developed radioautograph, silver grains identify cells with viral genes, and the brown DAB product marks oligodendrocytes. With this technique we can unambiguously identify the oligodendrocyte as one of the cellular targets for the inflammatory response to visna virus, and thus partly explain demyelination in this slow disease.

Host Antigens, Autoimmunity; Molecular Mimicry. One of the subtler forms of virus-induced neurological disease is an immunopathological process directed against "self" antigens. For example, a murine coronavirus (the JHM strain) induces a demyelinating disorder in rats that can be adoptively transferred with immune cells to uninfected recipients. The virus apparently sensitizes infected animals to myelin basic protein, and the demyelinating disease in the recipients is the result of a process analogous to allergic encephalomyelitis.

Viruses and cells also may have antigenic determinants in common. Some monoclonal antibodies to purified structural subcomponents of measles and herpes simplex viruses cross-react with constituents of normal cells. One antibody to the phosphoprotein of measles virus also recognizes an intermediate filament protein, probably vimentin. We anticipate autoimmune disorders that arise as the consequence of molecular mimicry of host antigens by the virus.

Viral Immunopathology and a Behavioral Disorder. Immunopathology in viral infections is not limited to demyelination. In Borna disease of rats the immune response to the as-yet-unclassified etiological agent is primarily in the limbic system, and infected animals display behavior somewhat analogous to schizophrenia. In later stages of infection the virus is not eliminated, but the inflammatory response subsides and so does behavioural disturbance.

Dysfunction

Virus infections may be the source of quite profound metabolic derangements without any evidence of cytopathology. Lymphocytic choriomeningitis virus produces runting in newborn mice by interfering with the elaboration of growth hormone. The persistently infected cells in the anterior pituitary are ostensibly normal in appearance, even at the electronmicroscopic level.

Unconventional Agents and Dementia

Infections also provide good models for aging and dementia. Scrapie is a filterable agent like a virus but with a number of unusual properties. Some strains of the agent produce pathological changes that resemble those in Alzheimer's disease, such as the neuritic plaques and disposition of amyloid; moreover, scrapie-associated fibrillar material exhibits the same staining properties as amyloid.

GENE EXPRESSION IN CHRONIC VIRUS INFECTION

While viruses and unconventional agents may provide interesting and informative infectious disease models for dementia, demyelination, and behavioural disorders, the relevance of these models to their human counterparts has yet to be established. One of the major objections to viruses as agents of chronic neurological disease is our lack of success in demonstrating viral antigens, viral particles, or transmissable infectivity in samples obtained from patients afflicted with Alzheimer's disease, MS, or schizophrenia. The thrust of the succeeding sections will be to demonstrate that this is the expected result in slow infections, because of restricted synthesis and expression of virus genes in chronic virological infections.

RESTRICTED SYNTHESIS AND EXPRESSION OF VIRAL GENES AS A UNIFYING THEME IN SLOW INFECTIONS

In slow infections viruses must survive for extended periods of time, often in the face of a vigorous humoral and cellular response mounted by the host; the host also must survive for years with viruses that kill cells in a culture in a matter of days. There is now considerable evidence that viruses in slow infections satisfy these logical requirements through a limitation in the synthesis and expression of their genomes. The block in expression of antigens provides an explanation for the inability of host immunity to eradicate the covertly infected cells, and the restriction in the synthesis of viral gene products mitigates the lytic effects of viral replication.

VISNA

This understanding of the molecular basis of slow infections was initially derived from analysis of the slow infection visna, a demyelinating disease of sheep. The etiological agent, a retrovirus, replicates and kills cells in tissue cultures in a few days, but in infected animals disease becomes manifest often only years after the introduction of the virus. Analyses of the small fraction of infected cells (typically 10^{-2} to 10^{-4} in the population) by in situ hybridization revealed a block in the synthesis and transcription of the viral genome in the animal. As a consequence, the infected cells contain too few antigens for detection and destruction by immune surveillance. However, an immunopathological response described earlier evolves slowly, commensurate with the gradual accumulation of antigens in some of the infected levels to levels that will provoke and sustain inflammatory disease.

SUBACUTE SCLEROSING PANCENCEPHALITIS (SSPE)

Similar observations and conclusions hold in SSPE, a rare complication of measles infection, that occurs many years after a primary infection from which the child apparently recovers. Early in the clinical course of SSPE, virus genes can be detected by in situ hybridization in about 0.1% of the cells, but there are fewer than ten copies of viral RNA per cell, and only a rare cell in the infected population has detectible antigen. By contrast, in permissive infections in tissue culture nearly all the cells in the population bear detectible antigens; each cell contains more than a thousand copies of viral RNA.

IMPLICATIONS: SURVEY FOR MEASLES VIRUS RNA IN MS

The implication of these findings is that previous attempts to demonstrate viral antigens in chronic neurological diseases predictably would have met with failure. The question of the role of viruses in chronic disease accordingly must be reopened, but with a search for viral genomes. This was the motive and rationale underpinning our reexamination of MS, a disease with an epidemiology consistent with a slow infection. In the past three years we have found evidence of measles virus genes in low copy numbers in the brains of patients succumbing to MS, but also in controls. Because the infected cells occur in both pathologically involved and normal tissue, there is no reason to assume that measles virus causes MS. However, there is also no reason to exclude a role for the virus, perhaps by inciting an immunopathological process in susceptible individuals.

NEW METHODS FOR THE DETECTION OF VIRAL GENES

We think that the deposition and persistence of viral genes in the human CNS is itself intriguing from the standpoint of the molecular ecology of virus infection. We therefore plan to continue this search for the common viruses, both to satisfy our curiosity and to determine what relationship, if any, there might be to neurological diseases. Because two new techniques devised for this search may prove equally valuable to neurobiologists, we will conclude by describing a method to detect genes in whole animals or organs and a method to detect two different genes in cells from the same tissue section.

Combined Microscopic-Macroscopic Hybridization

Very large frozen sections of whole animals or organs can be cut and transferred to clear tape or glass. Hybridizing with a probe labelled with ^{125}I and ^{35}S will produce a signal that can be detected with great sensitivity both at the microscopic and macroscopic level. The gamma irradiation of ^{125}I with intensifying screens will produce a record on X-ray film of the location of viral genes in the tissue. The darkened area of the film subsequently serves as a mask or guide to the region of the section that warrants microscopic examination, after coating the section with emulsion, developing it, and staining it. As matters stand currently, about 50 copies of visna RNA (101kb) can be detected on X-ray film. This assay greatly extends the range and power of hybridization for sampling large areas of tissue without sacrificing the fine cellular resolution of in situ hybridization.

Double Label In Situ Hybridization; Color Microradioautography

The genes of herpes simplex and herpes zoster viruses have been described in the human CNS as well as measles virus. To examine the cellular interrelationships of these viruses, we have invented a double label in situ hybridization method that in principle should be applicable to the analysis of two different genes in the same or different cells in a tissue section. The method is based on the different energies of ^{35}S and ^{3}H labelled probes and color microradioautography to facilitate distinguishing the two isotypes. After hybridizing with a ^{3}H probe for one kind of gene and a ^{35}S-labelled probe for another, silver grains in the first layer of emulsion are converted to a magenta color. A thin film is applied that blocks penetration of ^{3}H into a second layer of emulsion. After development, silver grains in the second layer are converted to a cyan color. Cells that bind ^{3}H have only magenta grains. Cells annealing both probes have an increased number of magenta grains in the first layer. The contribution of ^{35}S to the total number of magenta grains can be estimated by enumerating cyan grains.

Neurobiology: Molecular Biological Approaches to Understanding Neuronal Function and Development 127–134 (1985)

Regulation of Neuropeptide Biosynthesis in the Central Nervous System

Jeffrey F. McKelvy, Jeffrey D. White, Kim D. Stewart, James E. Krause, Jonathan A. Fuller, and Norbert E. Kremer

Department of Neurobiology and Behavior, SUNY at Stony Brook, Stony Brook, New York 11794

The importance of neuropeptide transmitters in the function of the nervous system has been increasingly investigated during the past two decades. Despite the large number of neuropeptides that are known to be synthesized in and released from neurons, the precise role of most neuropeptides in their projection systems is unknown. Furthermore, the regulation of the synthesis of neuropeptide precursor proteins and of the subsequent post-translational processing of these precursors is also unknown. In an effort to investigate this latter problem, we have begun studies of the biosynthesis and processing of hypothalamic neuropeptides in vivo. Using our procedures for the delivery of radiolabeled amino acids and the purification of radiolabeled neuropeptides, we have been able to draw two conclusions regarding the regulation of neuropeptide biosynthesis and processing for oxytocin and enkephalin: 1) the rate of oxytocin biosynthesis vs vasopressin biosynthesis is increased approximately 50% in the hypothalamoneurohypophysial system during lactation, and 2) the processing of the enkephalin precursor in the hypothalomohyophysial system changes during lactation. The implications of these results are discussed.

Key words: neuropeptides, biosynthesis and processing, central nervous system

The role of neuropeptides in neuronal function is now a subject of intensive research. The number of neuropeptides that have been isolated, sequenced, and localized in the nervous system has risen dramatically in the last several years (for a review see [1]). However, despite this large increase in the information available regarding the sequence and localization of neuropeptides, little is known about the precise role of most neuropeptides in defined projection systems within the nervous system. The goal of the research in our laboratory has been to address the role of neuropeptides from a functional biochemical point of view; that is, how do the biochemical processes operative in peptide-secreting neurons subserve the physiolog-

J.E. Krause's present address is Department of Anatomy and Neurobiology, Washington University School of Medicine, St. Louis, MO 63110.

Received May 15, 1984.

ical roles of neuropeptides? As a subset of this general question we have asked the following: 1) At what step(s) is neuropeptide biosynthesis regulated (transcriptional, translational, post-translational)? 2) Does projection-specific processing of a given neuropeptide precursor occur in the CNS? 3) Are co-localized neuropeptide genes coordinately or independently regulated? We have chosen to investigate these questions at both the in vivo and in vitro levels.

The recent demonstration that known neuropeptides are cleaved from larger precursor proteins (Fig. 1) suggests a polyprotein nature for these precursors [5]; ie, that several biological signals are encoded in these structures. The precursors for the family of opioid peptides (proopiomelanocortin [POMC], proenkephalin, and prodynorphin) may represent the best models for such high information systems. Several studies have demonstrated, eg, that the single POMC precursor will yield a variety of related peptides and that this differential cleavage is dependent upon the location of the cells under study, intermediate lobe vs anterior lobe of the pituitary [6]. In addition, proenkephalin and prodynorphin each contain several potential opioid peptides, which could be liberated by endoproteolysis of flanking paired basic or single arginine residues [4,7]. The possibility for further chemical complexity at synapses involving peptides is afforded by observations of the coexistence of neuropeptides with conventional neurotransmitters and with other neuropeptides, as has been demonstrated in several central nervous system (CNS) nuclei and in peripheral ganglia throughout the neuraxis [8]. This picture suggests the possibility for variability in the chemical output of peptide-secreting neurons. Such variability could comprise a kind of plasticity at the molecular level if it is coupled to different patterns of activity of a neuronal nucleus. Clearly, in order to investigate this question it is necessary to investigate both pre- and post-translational regulation of neuropeptide precursors in a defined physiological context.

As a model system for our studies of neuropeptide biosynthesis and processing, we have chosen hypothalamic neuroendocrine systems. These systems represent good model systems for a variety of reasons: 1) the projections of the neurons to the median eminence, the posterior lobe of the pituitary, and CNS sites are well defined; 2) the cells are amenable to physiological, as compared to pharmacological, stimulation; 3) the nuclei are small and discrete, which facilitates both protein biosynthetic studies and the harvesting of tissue for nucleic acid studies; and 4) the coexistence of neuropeptides within single neurons has been demonstrated for several hypothalamic neuropeptides.

Figure 1 illustrates the precursor structures of the three neuropeptides that will be discussed at length below. Oxytocin, vasopressin, and their associated neurophysins have long been known to be synthesized in the paraventricular and supraoptic nuclei (PVN and SON) of the hypothalamus and released from nerve terminals in the external zone of the median eminence and the neural lobe of the pituitary. In fact, the concept of neurosecretion was first developed in this system nearly 50 years ago by the Scharrers [9]. Vasopressin is secreted in response to osmotic stimuli to promote water reabsorption by the kidney, whereas oxytocin is specifically released during suckling to initiate milk ejection by the mammary epithelia. Recently, it has been demonstrated that opioid peptides coexist with the nonapeptides in this system; vasopressin coexists with dynorphin [10], and oxytocin coexists with methionine enkephalin-containing peptides [11,12]. Thus, using the lactating female rat as an experimental system, we can investigate two aspects of plasticity for neuropeptides

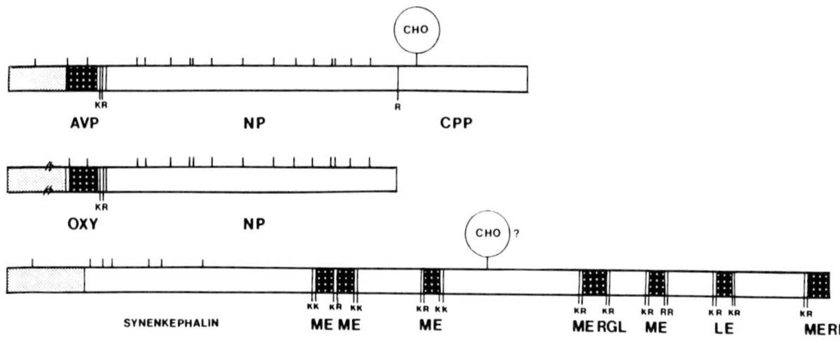

Fig. 1. Deduced structures of the precursors to AVP, OXY, and MENK. Shown are the structures of the precursor proteins to vasopressin-neurophysin, oxytocin-neurophysin, and polyenkephalin. These structures are predicted from the deduced amino acid sequences from cDNA cloning [2–4]. In all cases, the stippled areas denote the putative signal peptide, the darker areas are the mature peptides, R = arginine, K = lysine (presumed recognition sites for proteolytic processing), up slashes denote cysteine residues, and CHO = carbohydrate. NP = neurophysin and CPP = carboxy terminal glycopeptide of provasopressin. ME and LE are Met- and Leu-enkephalin respectively; MERGL and MERF are defined in the text.

that coexist: 1) increased biosynthesis in response to increased release (for oxytocin) and 2) plasticity with respect to the processing of a multi-functional precursor.

PROTEIN BIOSYNTHETIC STUDIES

We have previously described methods for the study of neuropeptide biosynthesis in vivo [13]. This method uses osmotic minipumps to deliver a pulse of radioactive amino acids through stereotaxically implanted stainless steel cannulae. We have successfully employed this method for a number of CNS projection systems, including LHRH biosynthesis in the preoptic area and transport to the median eminence [14] and substance P biosynthesis in the corpus striatum and transport to the substantia nigra [15]. When the biosynthesis and processing of POMC in the arcuate nucleus was investigated using this method, we found that the processing of the precursor more closely resembled that in the intermediate lobe of the pituitary. Thus, projection-specific processing of a multi-functional precursor was demonstrated for a CNS system in this way [16]. We have also used this in vivo system to investigate the biosynthesis of vasopressin (AVP), oxytocin (OXY), and their neurophysins in the PVN and the transport of these peptides to the posterior pituitary and to the nucleus of the solitary tract in the medulla [17]. These initial experiments with AVP and OXY have permitted us to develop sequential HPLC purification systems that result in the purification of the nonapeptides to constant specific activity and do not require the use of antibody affinity purification systems.

OXYTOCIN AND ENKEPHALIN BIOSYNTHESIS DURING LACTATION

As mentioned above, we have chosen the lactating female rat as a model system in which to investigate a number of questions regarding the regulation of neuropeptide biosynthesis and processing. It has been demonstrated by a number of investigators, most notably Lincoln, Wakerley, and coworkers (see [18] for a review), that the release of oxytocin from the posterior pituitary is specifically elicited by the suckling of pups during lactation. The normal firing pattern of vasopressin neurons is not altered during this stimulation. In the rat, the release of oxytocin from the posterior pituitary is initiated by the suckling stimulus, and this first release of peptide occurs after a brief lag period. Thereafter, the release is phasic, occurring as a burst of release at regular intervals [19]. Oxytocin released from the median eminence has also been suggested to play a role as a prolactin-releasing factor [20]. Hence, during lactation oxytocin may act at two sites to coordinate milk production and release. It is this specificity of response of the oxytocin neurons vs vasopressin-containing neurons that makes lactation an attractive model. This specificity allows us to compare the ratios of radioactivity present in the two nonapeptides in a given animal and then to compare these ratios between animals. The ability to compare ratios of radioactivity, instead of absolute amounts of radioactivity, allows us to circumvent the inherent limitations of the in vivo technique. These limitations include variability in the specific activity of the amino acid delivered, variable recovery of the amino acid to be delivered, and slight variations in cannulae placement. This rationale can also be used to investigate the biosynthesis of the Met-enkephalin peptides in this system. This makes it possible to ask experimentally if the biosynthesis of enkephalin, co-localized in PVN neurons secreting oxytocin, also changes in response to the stimulus of lactation. Such information is relevant to understanding how the activation of neuropeptide genes that are expressed in the same neuron may coordinate responses to physiological change. As Figure 1 illustrates, the three enkephalin peptides studied here, Met^5-enkephalin (MENK), Met^5-Arg^6-Gly^7-Leu^8-enkephalin (MERGL), and Met^5-Arg^6-Phe^7-enkephalin (MERF), are all derived from the same precursor. Thus, a comparison of the ^{35}S-methionine radioactivity present in each of these peptides allows us to investigate the post-translational processing of the polyenkephalin precursor by cells in the PVN.

The general experimental protocol for these experiments has been to infuse ^{35}S-amino acid above the PVN for either 2 hr or for 6 hr, then to allow a "chase" time for the amino acid to be incorporated into precursor, processed, and transported to the terminal fields. The median eminence and the posterior pituitary are harvested, the peptides extracted in the presence of unlabelled carrier peptides, then purified using sequential HPLC and the chemical modification, sulfoxidation, and sulfone formation of the peptides. Both standard organic elution from reverse phase columns and ion-pairing buffer systems are used to insure the peptides are purified to constant specific activity; ie, a constant ratio of radioactivity to integrated area of the carrier peptide peak.

In our preliminary experiments, we have been able to obtain data concerning the nonapeptides in the posterior pituitary and the enkephalins in the median eminence. As Table I indicates, the ratio of radioactivity in oxytocin to vasopressin suggests that oxytocin biosynthesis is stimulated approximately 1.5-fold in the lactating female when compared to the normal female. The observation of greater amounts of radioactivity present in oxytocin than in vasopressin at each of the times tested is further support for an increase in the rate of oxytocin biosynthesis. This finding has allowed us to conclude that lactation will be a good model for the study of the selective activation of oxytocin neurons. Our future experiments will be directed at the study of oxytocin and vasopressin in the median eminence. With the knowledge that the model is appropriate, we have also investigated the biosynthesis of the

TABLE I. Oxytocin and Vasopressin in the Neural Lobe*

Condition		^{35}S-DPM OXY/^{35}S-DPM AVP
Lactation		
Pulse	Chase	
2	2	1.52
2	4	1.52
6	2	1.60
Overall mean		1.54
Normal (diestrus)		
Pulse	Chase	
2	4	0.82
6	2	1.38
Overall mean		1.10

*Animals were stereotaxically cannulated, and the radioactive isotope was delivered as described in the text and elsewhere [13]. ^{35}S-OXY and -AVP were purified to constant specific activity as in [17]. The data are expressed as the ratio of radioactivity present in OXY vs the radioactivity present in AVP. Pulse and chase times are in hr.

enkephalins and have been able to draw several conclusions. Although there is sufficient radioactivity present in the median eminence for study, the level of radioactivity present in the neural lobe has been quite low. This finding is consistent with the notion that the oxytocin and enkephalin neurons of the PVN project either to the posterior pituitary or to the median eminence but that most of the enkephalin present in the neural lobe is derived from projections from the SON. We have been able to draw another, and rather striking conclusion based upon the data presented in Table II. In these experiments the ratio of radioactivity present in MENK, MERGL, and MERF was examined in the median eminence. In all cases, the amount of radioactivity present in MERF was arbitrarily set equal to one, and the data are presented as values relative to this value. If the post-translational processing of the enkephalin precursor was complete after all paired basic residues, then we would expect a ratio of MENK:MERGL:MERF of 4:1:1. When this ratio is examined in normal (diestrous) females, we then find a ratio of approximately 4.4:0.6:1. The simplest explanation for this ratio is that processing of MERGL takes place beyond the flanking basic residues. Naturally, a number of other processing events could result in this ratio. However, when this ratio is examined in the lactating female, a different ratio is observed. Following a 2-hr administration of labeled amino acid and a 4-hr period for processing and transport, the ratio of MENK:MERGL:MERF in the median eminence is 2.6:0.5:1. With only a 2-hr period for processing and transport, the ratio is 3.4:0.5:1. Clearly, these ratios are different from the ratios observed in the normal animal. One explanation for the difference in the ratios in the two chase periods is that MENK is preferentially cleaved first from the precursor, followed by MERF, then MERGL. Most importantly, however, these data are consistent with the hypothesis that under this controlled state of physiological activation of the PVN neurons there is regulation of the processing of the enkephalin precursor, resulting in a change in the peptides that are released from the median eminence.

TABLE II. Enkephalin Ratios in the Median Eminence*

Condition		MENK:MERGL:MERF
Lactation		
Pulse	Chase	
2	2	3.4:0.5:1
2	4	2.6:0.6:1
Normal		
Pulse	Chase	
2	4	4.4:0.6:1
6	2	4.4:0.7:1

*Animals were cannulated, and the isotope was delivered as above. The data are expressed as the ratios of ^{35}S-Met DPM in each of the peptides, with the amount of radioactivity present in MERF set equal to one. Pulse and chase times are hr.

NUCLEIC ACID STUDIES

Based upon the success of the protein biosynthetic studies, we have also initiated studies to examine changes in the mRNA population that code for pro-AVP, pro-OXY, and proenkephalin. In these initial experiments we have purified total RNA from dissected hypothalami from lactating and normal female rats. These RNAs were then dotted onto nitrocellulose paper and allowed to hybridize either with ^{32}P-labeled synthetic oligonucleotide probes complementary to the known coding sequence for AVP or for OXY or with a ^{32}P-labeled restriction endonuclease fragment from the cDNA clone to human proenkephalin (generously provided by Dr. Edward Herbert). To date, these hybridization experiments have supported the conclusions from the protein biosynthetic studies. There appears to be an increase in the amount of material that hybridizes to the OXY probe in the RNA preparations from lactating animals when compared to the hybridization to the RNA from normal animals. Similarly, there appears to be an increase in the amount of material that hybridizes to the enkephalin cDNA in the lactating vs the normal animal (see Fig. 2).

FUTURE DIRECTIONS

The results we have cited above clearly demonstrate the need to investigate the regulation of the production of neuropeptides by neurons during physiological stimulation. The ability to study the regulation of specific mRNAs in the nervous system will afford the possibility to separate short-term regulation from long-term regulation. In all likelihood, studies such as these will lead to the examination of the activation and regulation of gene transcription in the nervous system and, therefore, to a more complete understanding of processes such as behavior and memory. However, despite the revolution in molecular biology and its application to the nervous system, it is essential to remember that information processing and information exchange in the nervous system takes place at the protein level. Thus, studies such as those described above for the proenkephalin peptides are essential to an understanding of communication in the nervous system. Indeed, it is these studies of the regulation of peptide

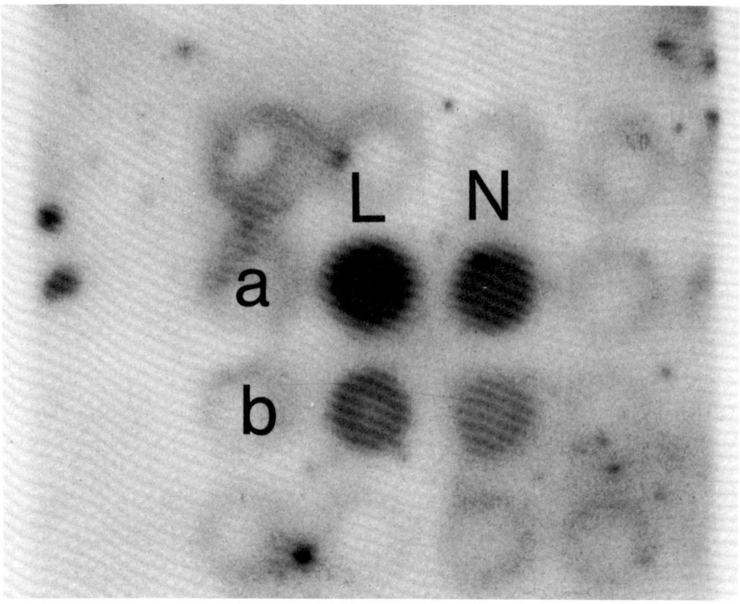

Fig. 2. RNA dot hybridization for proenkephalin RNA. Shown is an autoradiogram from an experiment in which two different concentrations of total RNA from whole hypothalamus was dotted onto nitrocellulose paper and hybridized with a ^{32}P-labeled restriction endonuclease fragment from a cDNA to human proenkephalin [4]. RNA was prepared from normal females (lane N) and from lactating females (lane L). Hybridization was overnight at 37°, and the final wash conditions were 0.1 × SSC/1%SDS at 55°. RNA concentrations dotted in rows a and b were: 9 μg and 1 μg for L and 8 μg and 2 μg for N, respectively. Note the stronger signal from the lactating animals.

processing that may lead to a fundamental change in the understanding of the nature of communication in the nervous system. Specifically, these studies may demonstrate that a given neuron can change not only the quantity of transmitters that it releases (eg, through an alteration in the firing pattern of the neuron) but also the quality of the transmitters it releases, and thereby communicate different information based upon a different physiological status.

REFERENCES

1. Krieger DT: Science 222:975, 1983.
2. Land H, Schutz G, Schmale H, Richter D: Nature 295:299, 1982.
3. Land H, Grez M, Ruppert S, Schmale H, Rehbein M, Richter D, Schutz G: Nature 302:342, 1983.
4. Comb M, Seeburg PH, Adelman J, Eiden L, Herbert E: Nature 295:663, 1982.
5. Herbert E, Uhler M: Cell 30:1, 1982.
6. Rosa PA, Policastro P, Herbert E: J Exp Biol 89:215, 1980.
7. Kakidani H, Furutani Y, Takahashi H, Noda M, Morimoto Y, Hirose T, Asai M, Inayama S, Nakanishi S, Numa S: Nature 298:245, 1982.
8. Chan-Palay V, Palay S: "Co-existence of Neuroactive Substances in Neurons." New York: J. Wiley and Sons, 1983.
9. Scharrer E, Scharrer B: Biol Rev 12:185, 1937.

10. Watson SJ, Akil H, Fischli W, Goldstein A, Zimmerman E, Nilaver G, van Wimersma Greidanus TB: Science 216:85, 1982.
11. Martin R, Geis R, Holl R, Schafer M, Voight HK: Neurosci. 8:213, 1983.
12. Vanderhaeghen JJ, Lotstra F, Liston DR, Rossier J: Proc Natl Acad Sci USA 80:5139, 1983.
13. McKelvy JF, Krause JE, White JD: In Conn PM (ed): "Methods in Enzymology, Vol. 103, Neuroendocrine Peptides." New York: Academic Press, 1983, pp 511–523.
14. Krause JE, Advis JP, McKelvy JF: Endo 111:344, 1982.
15. Krause JE, Reiner AJ, Advis JP, McKelvy JF: J Neurosci 4:775, 1984.
16. Liotta AS, Advis JP, Krause JE, McKelvy JF, Krieger DT: J Neurosci 4:956–965, 1984.
17. White JD, Krause JE, McKelvy JF: J Neurosci 4:1262–1270, 1984.
18. Poulain DA, Wakerley JB: Neurosci 7:773, 1982.
19. Lincoln DW, Hill A, Wakerley JB: J Endocr 57:459, 1973.
20. Lumpkin MD, Samson WK, McCann SM: Endo 112:1711, 1983.

Neurobiology: Molecular Biological Approaches to Understanding Neuronal Function and Development 135–143 (1985)

Selective Regulation of Two Axonal Cytoskeletal Networks in Dorsal Root Ganglion Cells

Monica M. Oblinger and Raymond J. Lasek

Department of Developmental Genetics and Anatomy, Case Western Reserve University School of Medicine, Cleveland, Ohio 44106

Vertebrate neurons elaborate two very different kinds of cellular extensions, dendrites and axons. Dendrites are specialized for receptor functions, and axons are specialized for effector activities. These two types of regions of the neuron differ dramatically in external form, and this is related to detailed differences in their cytoskeletons [1–5]. An important question in cellular neurobiology is how are these regional differences in the cytoskeleton of neurons generated and maintained?

Our understanding of the mechanisms that produce regional differences in neurons has benefited principally from studies on an unusual neuron, the vertebrate dorsal root ganglion cell. Although these neurons do not have dendrites, they permit detailed analyses of the dynamic mechanisms that generate regional differences in two sets of processes that originate from the same neuron.

TWO AXONS OF DORSAL ROOT GANGLION CELLS: SPECIAL EXAMPLES OF SEGREGATED CYTOSKELETAL NETWORKS IN NEURONS

The dorsal root ganglion (DRG) cell is well suited as a model for the study of various parameters involved in the generation of regionally differentiated cytoskeletons in neurons. The mature DRG cell has two long axons, which branch from a short stem axon (Fig. 1). One axon courses centrally through the dorsal root to synapse on neurons in the central nervous system, the other proceeds peripherally to contact sensory receptors. Comparisons of the two axons in the adult have shown that the diameter of the peripheral axon is larger than that of the central axon [6], and ultrastructural studies have shown that the ratio of microtubules to neurofilaments is significantly greater in the peripheral than in the central axon [7,8].

The developmental history of the DRG cell provides clues to the marked differences in the structure of the central and peripheral axons of these cells. During their initial development, these two axons develop separately from the poles of the

Received May 21, 1984.

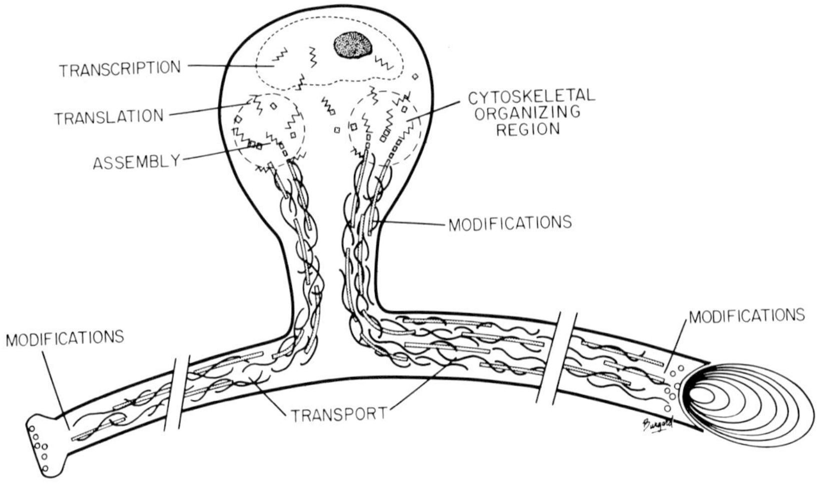

Fig. 1. Schematic of the organizing regions for the two axonal cytoskeletal networks in dorsal root ganglion cells. A summary of the major steps involved in the generation of the axonal cytoskeleton in each axon is depicted. The cytoskeletal proteins are synthesized in the cell body, assembled into cytoskeletal polymers, and transported into each axon of the DRG cell. Newly synthesized cytoskeletal proteins are assembled into the cytoskeletal network at the organizing region. The cytoskeletal components are transported with the cytoskeletal network to the end of the axon, where they are degraded by proteolytic mechanisms operating specifically in the axon terminal.

bipolar embryonic DRG cell [9,10]. Then, during subsequent development of the DRG cell, the two axons become apposed to form a single stem axon that bifurcates. Apparently, the embryonic bipolar neuron generates two separate cytoskeletal networks, one for each axon. Studies of the transport of the cytoskeletal networks in the central and peripheral axons of mature DRG cells indicate that the networks remain segregated after the axons have fused together [11].

Axonal cytoskeletal networks are actively transported from the neuron cell body toward the axon terminal by slow axonal transport and studies of this process have played an important part in understanding the metabolic pathway by which axonal cytoskeletons are generated and continually renewed. Axonal cytoskeletons originate in the neuron cell body, where the cytoskeletal proteins are synthesized [12]. The cytoskeletal proteins are assembled into polymers and other cytoskeletal structures near the site of synthesis and are then conveyed into the axon as component parts of the cytoskeletal network [11,29]. The cytoskeletal networks move unidirectionally from the cell body to the end of the axon, where they are degraded [13,14].

Slow axonal transport has been divided into two subcomponents, SCa and SCb [14–18]. Figure 2 illustrates the polypeptide composition of these components in the central and peripheral axons of dorsal root ganglion cells. The SCa component corresponds to the microtubule-neurofilament network and consists of neurofilament proteins, tubulin, and associated proteins. The SCb component contains two cytoskeletal proteins, actin and tubulin, as well as many proteins that constitute the cytoplasmic matrix. The amount and rate of transport of cytoskeletal proteins conveyed by SCa and SCb are important determinants of axonal structure. Comparisons of slow

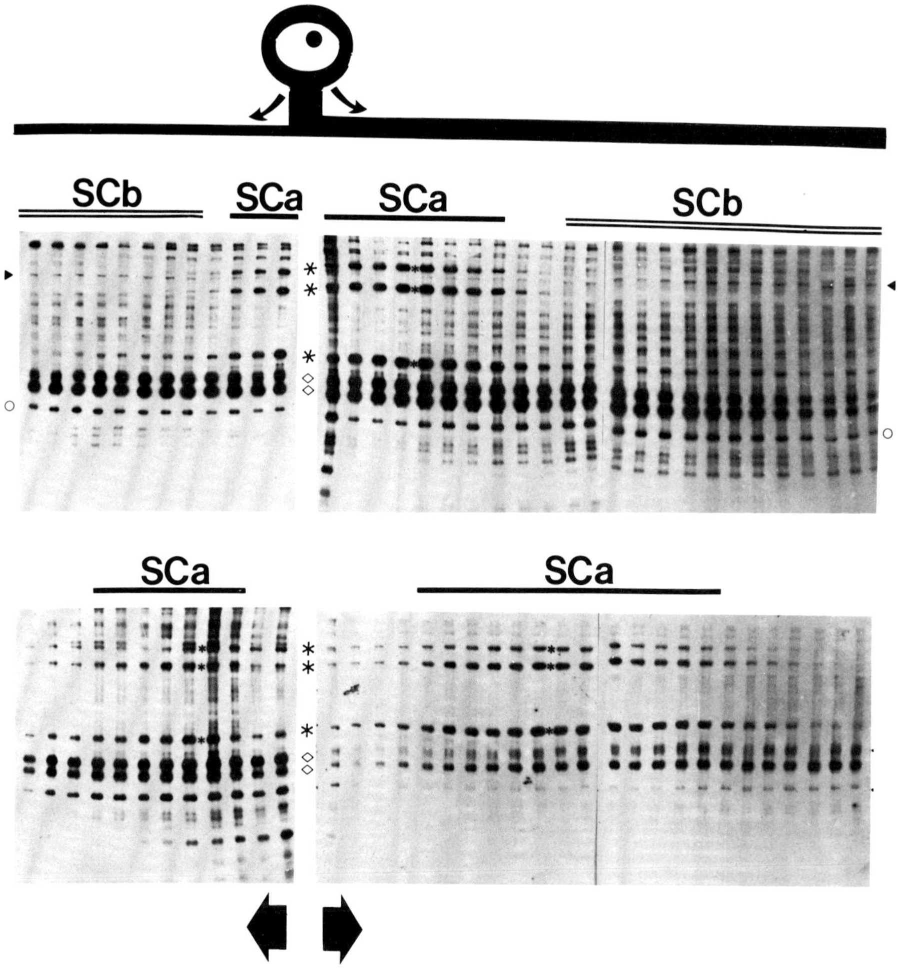

Fig. 2. Fluorographs illustrating differences in the rate and amount of the two components of slow axonal transport in the two axons of dorsal root ganglion cells. Each lane in the fluorographs consists of consecutive 2-mm segments of central branch (left) or peripheral branch (right) axons of DRG cells in adult rats. In the top panel, axons were retrieved seven days after injecting ^{35}S-methionine into the L5 DRG; in the bottom panel, a 14-day interval elapsed prior to sacrificing the animal. The major proteins that constitute the SCa wave of slow transport are the neurofilament proteins (asterisks) at 200, 145, and 68 kilodaltons, and tubulin (open diamonds). The advance of the SCa wave in axons that occurs over one week can be appreciated by observing the location of the peak of the neurofilament radioactivity (indicated by small asterisks within the lanes) at two times. The faster moving SCb wave is more complex in its composition, but three identified proteins are indicated. These include clathrin (arrowhead), actin (open circle), and tubulin (open diamonds).

transport in the peripheral and central axons of DRG cells demonstrate major differences between the rate and amount of both SCa and SCb in these axons.

The amount of slowly transported protein exported to the peripheral axon is greater than that shipped centrally (Fig. 2 [19,20]). For neurofilament proteins, the peripheral to central ratio is 2:1, and for tubulin the peripheral to central ratio is approximately 5:1 [19,21]. In addition to these differences in amount of protein transported, the rates of both SCa and SCb also differ. The rates of SCa and SCb in peripheral axons are 1 mm/day and 4 mm/day, respectively. These rates in the central axon are 0.4 mm/day and 2 mm/day (Fig. 2, [19,20]). Similarly, the rates of regeneration of the peripheral and central axons differ by a factor of two. The differences in the rates of regeneration of the two axons can be explained by the differences in the rate of slow transport, because slow transport is a rate-limiting factor in axonal regeneration [20]. The differences in the rate and amount of slow transport may also produce the differences in the diameters of the peripheral and central axons [21].

ORGANIZING REGIONS FOR AXONAL CYTOSKELETONS

The differences between the axonal cytoskeletons may arise initially in the nerve cell body, where the proteins are first assembled into cytoskeletal elements. Indeed, it has been proposed that the cytoskeletal network of each axon originates from a distinct region of the cell body—a cytoskeletal organizing region. Cytoskeletal organizing regions are those locations in neurons where newly synthesized proteins are initially added to the cytoskeleton (Fig. 1). The properties of the axonal cytoskeletal networks are first established at the cytoskeletal organizing regions. These organizing regions represent a critical control point, which regulates the amount of protein that is added to the cytoskeletal network.

The differences that have been observed in the rates and amounts of cytoskeletal protein transport in the central and peripheral axons suggest that there are basic differences between the cytoskeletal organizing regions of the DRG cell. The amount of protein that is added to the organizing region for the peripheral cytoskeletal network must be larger than that for the central network, because more SCa and SCb proteins are transported with the cytoskeletal network of the peripheral axon [7,8,19,21]. Two parameters determine the amount of assembly at a cytoskeletal organizing region. One parameter is the number of assembly sites for cytoskeletal polymerization. The other is the rate of polymerization at each assembly site, which is influenced by many factors, including the availability of monomer and cofactors necessary for polymerization.

The peripheral axon contains more microtubules and neurofilaments than the central axon, and the diameters of the two axons differ accordingly. These differences are most easily explained if the organizing center for the peripheral axon has more assembly sites for cytoskeletal polymers than the central organizing center. A difference in the number of assembly sites contained in each organizing region may first appear during the early stage of DRG cell development, when the axons first develop. For example, Tennyson [9] has observed differences in the amount of membranous systems that extend into each developing neurite in embryonic bipolar DRG cells.

THE AXONAL CYTOSKELETAL NETWORKS CAN BE SEPARATELY REGULATED IN MATURE DRG CELLS

Neurons continue to maintain the capacity to regulate the export of cytoskeletal proteins into the axon, even in the adult organism. For example, when neurons

TABLE I. Effects of Axotomy on the Amount of Cytoskeletal Proteins That Are Slowly Transported in Central and Peripheral Axons of Dorsal Root Ganglion Cells*

	Central axotomy		Peripheral axotomy	
	Peripheral axon	Central axon	Peripheral axon	Central axon
SCa neurofilament protein	—	—	↓	—
SCa tubulin	—	—	—	—
SCb tubulin	—	—	↑	—

*Direction of change in amount of these proteins is indicated. Dash indicates no change from normal; arrows indicate increase or decrease.

regenerate their axons, they may increase or decrease the amount of cytoskeletal protein that is assembled and transported into the axon [22,23]. An interesting question arises in the case of the DRG cell, because it has two cytoskeletal organizing regions. Can the DRG cell selectively alter the export of cytoskeletal proteins into one of its axons if only a single axon is damaged, or is there a coordinate response that effects both axons? In the following section we examine this question.

To determine whether cytoskeletal protein export into each DRG cell axon is altered coordinately or not during the regeneration of only one axon, we lesioned either the central or the peripheral axon and analyzed slow transport of cytoskeletal proteins into both axons. Specifically, either the fifth lumbar (L5) dorsal root was crushed near its entry into the spinal cord (30–32 mm from the ganglion) or the sciatic nerve was crushed at the knee (60 mm from the ganglion) in adult rats. This permitted the analysis of axonal transport in the region of the axon between the cell body and the lesion. Two weeks after crushing the axons, the L5 ganglion was labeled by injection of ^{35}S-methionine. Between 5–21 days after labeling the animals were sacrificed, and the slowly transported cytoskeletal proteins were analyzed by SDS-PAGE and fluorography.

Table I summarizes the results obtained in this study. Axotomy of only the central axon had no effects on the amount of slowly transported cytoskeletal proteins in either the central or peripheral axons (Fig. 3). This is consistent with studies of chromatolysis in DRG cells. When the central axon is lesioned, chromatolysis does not occur in these cells. By contrast, if the peripheral axon is lesioned, then the DRG cells exhibit a chromatolytic response, suggesting that these cells alter their protein metabolism under these conditions [24,25].

In fact, axotomy of only the peripheral axon of DRG cells resulted in two major changes in slow axonal transport (Table 1, Fig. 3). These occured in the cytoskeletal network of only the peripheral axon. One change was a decrease in the amount of labeled neurofilament protein transported in the peripheral axon. This is illustrated in fluorographs of slowly transported proteins shown in Figure 3 and has been confirmed by quantitative analysis (Oblinger and Lasek, unpublished). The other change was an increase in the amount of tubulin transported in SCb (Table 1, Fig. 3). Whereas peripheral axotomy stimulated the DRG cells to increase the export of SCb tubulin into peripheral axons, the amount of SCa tubulin did not appear to change. SCa and

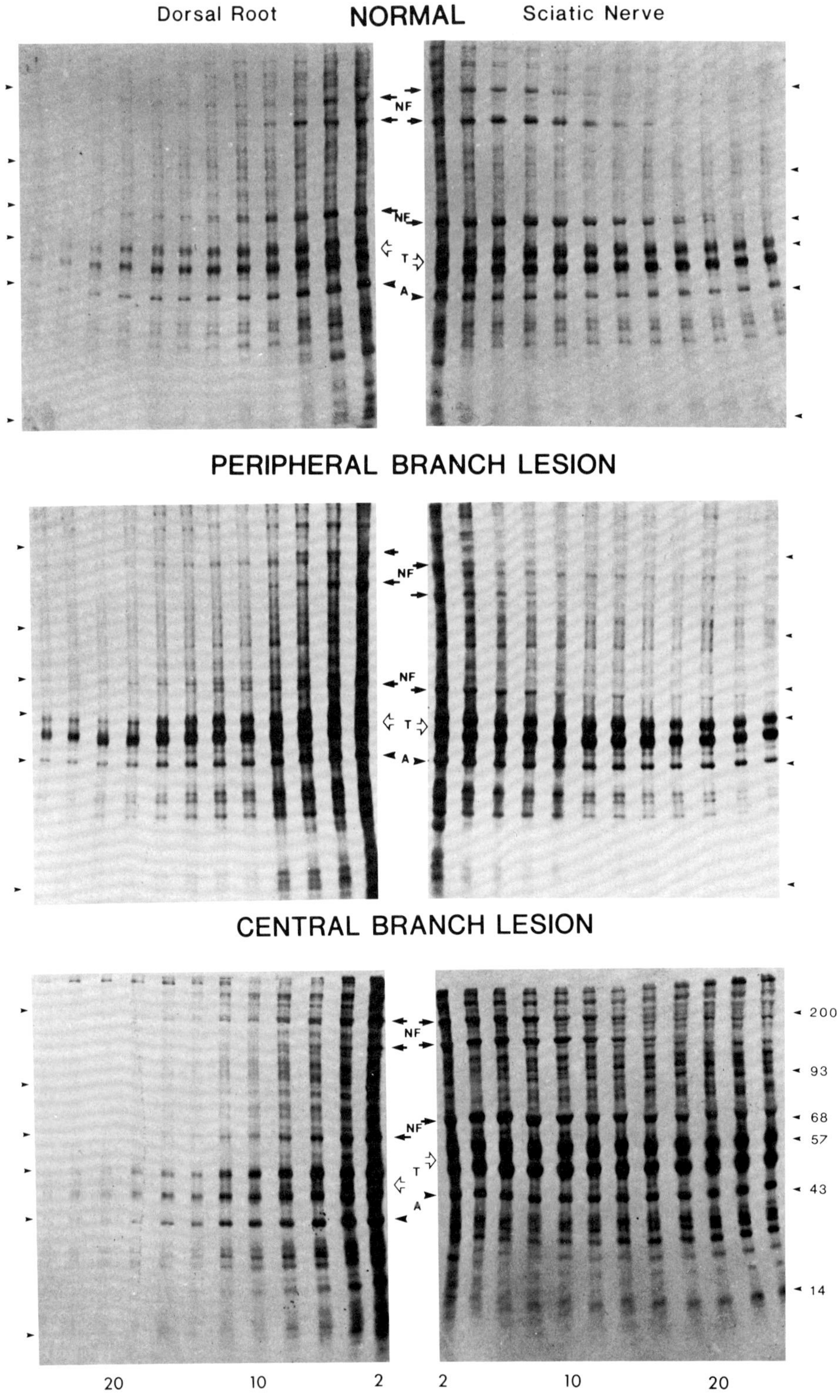
Dorsal Root
NORMAL
Sciatic Nerve
NF
NF
T
A
PERIPHERAL BRANCH LESION
NF
NF
T
A
CENTRAL BRANCH LESION
NF
NF
T
A
200
93
68
57
43
14
20
10
2
2
10
20

Fig. 3

SCb tubulin may have different isoforms [30]. Our analysis was not designed to separate these isoforms, and it will be interesting to determine whether there are selective changes in the isoforms of tubulin transported in the axon during regeneration.

These results demonstrate that the amount of neurofilament protein and SCb tubulin transported from the cell body into the peripheral axons can change without a concordant change in the central axon. Furthermore, the regulation of cytoskeletal polymers can be selective even in the same axon. We found that the amount of neurofilament protein decreased in the peripheral process but that the amount of tubulin specifically transported with SCb increased.

Changes in the amount of neurofilament protein and tubulin apparently correspond to selective changes in the number of neurofilaments and microtubules that are assembled and transported into the axon [29]. To explain these results. we propose that the organizing region for the peripheral axon is selectively altered during the regeneration of that axon. How might this organizing region change during regeneration? In order to produce a change in the number of microtubles and neurofilaments transported into the axon, there must be a change in the number of polymers assembled. The DRG cell might produce this change by varying the synthesis of protein monomers and regulating the amount of monomer available for assembly. In the next section we explore this possibility.

CYTOSKELETAL PROTEIN SYNTHESIS IN REGENERATING DRG CELLS

We have examined the synthesis of cytoskeletal proteins at various times after axotomy of the peripheral axon. Specifically, L5 DRGs were removed at 1, 5, and 14 days after sciatic nerve crush and incubated in sterile media containing ^{35}S-methionine at 37°C in 95% O_2 for one hour. Newly synthesized proteins were analyzed by 2D-PAGE/fluorography. Analysis of fluorographs of whole DRG synthesis showed that at five days after axotomy the incorporation of ^{35}S-methionine into neurofilament proteins was substantially lower than it was in control ganglia (Fig. 4). By 14 days after peripheral axotomy, this reduction was more prominent. By contrast, incorporation of ^{35}S-methionine into tubulin increased during this same period (Fig. 4). Preliminary quantitation of the individual proteins supports the qualitative observations made from fluorographs.

Fig. 3. Fluorographs of slowly transported proteins in normal and regenerating DRG axons. Gel lanes correspond to samples from consecutive 2-mm segments of either central branch axons in the dorsal root (left panesl) or peripheral branch axons in the sciatic nerve (right panels). The distance in mm from the L5 DRG is noted at the bottom of the figure. Normal axons were obtained seven days after injection of ^{35}S-methionine into the L5 DRG. Peripheral branch lesion data are from DRG cells subjected to a lesion of the sciatic nerve 55 mm from the DRG 2 wk prior to injection of isotope into the L5 DRG. Data from DRG cells that sustained a central axotomy were obtained from an animal that received a lesion of the L5 dorsal root 32 mm from the L5 DRG 2 wk prior to labeling the ganglion. In all cases, nerves were harvested at seven days after labeling the L5 DRG. A notable change in the slow transport profiles is a reduction in the amount of labeled neurofilament proteins (NF) that have entered the peripheral DRG axons (sciatic nerve) in response to previous peripheral axotomy. However, a similar reduction in labeled neurofilament protein is not observed in the central DRG axons in the dorsal root of the same animal. Central branch axotomy does not change the slow transport profiles in either the central or peripheral branch axons. T, tubulin; A, actin. Molecular weights (kilodaltons) are denoted by arrows to the side of each panel.

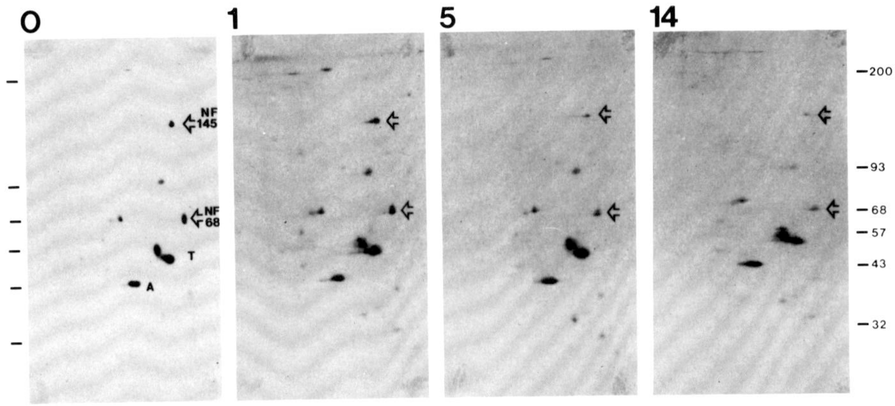

Fig. 4. Fluorographs of newly synthesized proteins in L5 DRG at various times after axotomy of the peripheral axon in the sciatic nerve. The fluorograph labeled 0 is from a normal DRG; other panels are from ganglia labeled at 1, 5, or 14 days after axotomy. At the indicated times the ganglia were removed, pulse-labeled with ^{35}S-methionine for 1 hr, homogenized, and the proteins were separated by 2D-PAGE. These fluorographs are from gels loaded with an equal amount of radioactive protein and are lightly exposed to illustrate only the major newly synthesized proteins in the DRG. The arrows indicate the position of two of the neurofilament proteins (NF). T, tubulin; A, actin. Molecular weight standards (kilodaltons) are indicated by numbers to the right side of the figure.

These changes in incorporation of labeled precursors into neurofilament protein and tubulin in regenerating DRG cells indicate that protein synthesis may be altered in response to peripheral axotomy. This possibility is supported by previous studies of tubulin synthesis in regenerating DRG cells and goldfish retinal ganglion cells [26–28]. Our observation that neurofilament protein labeling decreases suggests that changes in synthesis of these proteins contribute to the selective alterations in neurofilament protein transport that occur in response to peripheral axotomy. Similarly, increased tubulin synthesis may contribute to the selective increase of SCb tubulin that is transported into the peripheral axon in response to peripheral axotomy.

SUMMARY AND CONCLUSIONS

The DRG cell is a valuable model for studying the mechanisms that generate, maintain, and modulate the differences between different regions of the neuronal cytoskeleton. A variety of evidence suggests that the cytoskeletal networks of the peripheral and the central axons of the DRG cell originate from separate organizing regions in the cell body. These separate organizing regions apparently arise during the early bipolar phase of DRG cell development and remain separated throughout development and maturation. One manifestation of this separation is that the DRG cell can selectively alter the amount of protein that is added to one organizing center without changing the other one. In this way, the two cytoskeletal networks can be separately regulated in the mature DRG cell.

There are two ways by which the DRG cell could selectively modulate the amount of protein assembled at only one organizing center. One way is to selectively regulate the supply of protein for polymerization by locally varying such processes as protein synthesis. Another way is to vary the rate of assembly or the number of assembly sites at the organizing center. We have found that protein synthesis appears to change when the DRG cell selectively modulates the amount of protein exported into the peripheral axon. If this change in protein synthesis is selective for the peripheral organizing center, then this may be the point in the pathway at which the functional separation of organizing centers occurs.

REFERENCES

1. Peters A, Palay SL, Webster, HF: "The Fine Structure of the Nervous System." Philadelphia: W.B. Saunders, 1976.
2. Hillman DE: In Schmitt FO, Worden FG (eds): "The Neurosciences, 4th Study Program." Cambridge, Massachusetts MIT Press, 1980, pp 477–498.
3. Huber G, Matus A: J Neurosci 4:151, 1984.
4. Matus A, Bernhardt R, Hugh-Jones T: Proc Natl Acad Sci USA 78:3010, 1981.
5. Sternberger LA, Sternberger NH: Proc Natl Acad Sci USA 80:6126, 1983.
6. Suh YS, Chung K, Coggeshall RE: J Comp Neurol 222:473, 1984.
7. Smith RS: Can J Physiol Pharmacol 51:798, 1973.
8. Zenker W, Mayr R, Gruber H: Experientia 31:318, 1975.
9. Tennyson VM: J Comp Neurol 124:267, 1965.
10. Pannese E: Adv Anat Embryol Cell Biol 47.5:1, 1974.
11. Lasek RJ: Neurosci Res Prog Bull 19:7, 1981.
12. Lasek RJ, Brady ST: Cold Spring Harbor Symp Quant Biol 46:113, 1982.
13. Tytell M, Black MM, Garner JA, Lasek RJ: Science 214:179, 1981.
14. Lasek RJ: Trends Neurosci 3:87, 1980.
15. Brady ST, Lasek RJ: In Weiss DG (ed): "Axoplasmic Transport." New York: Springer-Verlag, 1982, pp 207–217.
16. Brady ST, Lasek RJ: In Wilson L (ed): "Methods in Cell Biology, Vol 25." New York: Academic Press, 1982, pp 365–398.
17. Grafstein B, Forman DS: Physiol Rev 60:1167, 1980.
18. Black MM, Lasek RJ: J Cell Biol 86:616, 1980.
19. Mori H, Komiya Y, Kurokowa M: J Cell Biol 82:174, 1979.
20. Wujek JR, Lasek RJ: J Neurosci 3:243, 1983.
21. Lasek RJ, Oblinger MM, Drake PF: Cold Spring Harbor Symp Quant Biol 48:731, 1983.
22. Hoffman PN, Lasek RJ: Brain Res 202:317, 1980.
23. McQuarrie IG: In Seil FJ (ed): "Nerve, Organ and Tissue Regeneration: Research Perspectives." New York: Academic Press, 1983, pp 51–86.
24. Lieberman AR: Int Rev Neurobiol 14:49, 1971.
25. Oblinger MM, Lasek RJ: J Neurosci 4:1736, 1984.
26. Burrell HR, Heacock AM, Water RD, Agranoff BW: Brain Res 168:628, 1979.
27. Hall ME: Exp Neurol 76:83, 1982.
28. Perry GW, Wilson DL: J Neurochem 37:1203, 1981.
29. Lasek RJ, Garner JA, Brady ST: J Cell Biol 99:212S, 1984.
30. Tashiro T, Kurokawa M, Komiya Y: J Neurochem 43:1220, 1984.

Neurobiology: Molecular Biological Approaches to
Understanding Neuronal Function and Development 145–149 (1985)

PDGF-Like Molecules: Multiple Sources and Roles

Daniel F. Bowen-Pope, Ronald A. Seifert, and Russell Ross

Department of Pathology, SM-30, University of Washington, Seattle, Washington 98195

Platelet-derived growth factor (PDGF) is one of an expanding list of hormones the existence of which was deduced from experiments performed using cells cultured ex vivo. Most of our hypotheses about the possible functions of PDGF in vivo still reflect the results of cell culture experiments. During the last two years, experiments with cultured cells have shown that PDGF-like molecules can be secreted by several cell types. These results, and their implications for the possible physiological roles of PDGF in vivo, will be discussed.

PDGF FROM PLATELETS

PDGF was discovered as a consequence of its presence, in high concentration, within human platelets [1–4]. PDGF from this source has been purified and shown to be mitogenically active in stimulating the proliferation and biosynthetic activity of a variety of cultured connective tissue cells, including vascular smooth muscle cells [reviewed in 5]. Ross and Glomset [6] have proposed that if vascular smooth muscle cells respond to PDGF in vivo as they do in culture, then some of the cell proliferation that is seen at sites of vascular damage may result from stimulation by PDGF released from platelets that degranulate in these areas in response to contact with the exposed connective tissue cells or matrix.

In the terminology proposed by Dockray [7] and Sporn and Todaro [8], the PDGF released from these platelets would be acting as a "paracrine" hormone, in that the source and target are in contact or in close proximity. The platelet is unusual in this regard, in that, since it circulates freely in the blood and can gain access to exposed (injured) extravascular tissues, it can act paracrine to almost any cell in the body, regardless of that cell's normal neighbors. This suggests one possible explanation for the finding that platelets contain many distinct growth factors, including PDGF [1–4], epidermal growth factor-like molecules [9,10], class beta transforming growth factor [11,12], and probably others [13]. Since different cell types respond to different mitogens, it may be necessary for the platelet to contain multiple growth factors to ensure that all appropriate cells are stimulated. For example, it is known that EGF [reviewed in 14] and PDGF [reviewed in 5] both stimulate connective tissue cell types, but that EGF can also act on epithelial cell types.

Received July 16, 1984.

An alternative reason for the presence of multiple growth factors in platelets may be related to possible differences in the growth requirements of cells in vivo compared to the in vitro environment. The so called "transforming growth factors" are assayed in vitro by their ability to stimulate growth of cells suspended in soft agar. "Mitogens" are assayed by their ability to stimulate growth of adherent cells. The relationship of cells in vivo to each other and to the more or less "solid" substrata that surrounds or borders them is varied and variable. It is not known whether fibroblasts in the dermis, or vascular smooth muscle cells in the multilayered aortic tunica media, behave in response to growth factors more like cells attached to plastic or like cells suspended in soft agar. In addition, the pattern of responsiveness also may be altered when a tissue is "disorganized" by being damaged. It is thus possible that many of the cells in vivo, the biosynthetic and proliferative rates of which have been proposed to be stimulated by "mitogens" released from platelets, are not in an environment that allows them to respond to this class of growth factor alone, and that additional factors are required to stimulate cell growth during wound healing [15]. The synergistic effects of EGF and beta-TGF, in stimulating the growth of cells in suspension [16,17], may thus reflect their physiological roles in vivo, ie, to stimulate cell proliferation in damaged tissue. Assoian et al [18] have reported that PDGF, which by itself does not promote the growth of cells in soft agar, may be synergistic with both EGF and beta-TGF in promoting growth in suspension. Therefore, an alternative hypothesis for the various growth factors contained in platelets is that they exert different and complementary effects on target cells, which would not respond to the individual factors alone.

PDGF-c FROM ENDOTHELIAL CELLS

In addition to PDGF released from platelets, there is now evidence that non-platelet sources of PDGF-like molecules exist and may play roles in stimulating cell proliferation in a number of circumstances. These PDGF-like molecules have been referred to as "PDGF-c" [19], in that they are assayed using their ability to compete with radioiodinated purified PDGF for binding to PDGF receptors on test cells. The first recognized source of PDGF-c from nontransformed cells was vascular endothelial cells. Medium conditioned by these cells had been previously shown to be mitogenic for smooth muscle cells and fibroblasts [20,21]. Subsequent analysis showed that this conditioned medium contains significant quantities of a molecule that competes with ^{125}I-PDGF for binding to the PDGF receptor on test cells [22]. The effects of both PDGF and endothelial cell PDGF-c on subsequent ^{125}I-PDGF binding could be reversed by incubating the test cells with acidified saline—suggesting that the mechanism of inhibition in both cases was reversible occupation of the receptor, rather than an irreversible (eg, proteolytic) effect.

Endothelial cell PDGF-c is quite similar to platelet PDGF when compared by several criteria. All the PDGF-c activity of endothelial cell-conditioned medium can be neutralized by incubation with an antiserum raised against purified PDGF. In addition, both endothelial cell PDGF-c and platelet PDGF are 30–50,000 molecular weight proteins that are stable to boiling but are inactivated by reduction [23]. It is not yet known whether PDGF and endothelial PDGF-c are products of the same gene and, if so, whether they are processed in identical or somewhat different fashions by the two cell types.

Not all of the mitogenic activity of endothelial cell-conditioned medium can be accounted for by PDGF-c. At least 75% of the mitogenic activity of endothelial cell-conditioned medium is not neutralized by antiserum against PDGF and can be separated from PDGF-c by cation exchange chromatography [23]. Thus, endothelial cells resemble platelets, in that they can release a battery of different growth factors—only one of which is PDGF-like.

PDGF-c FROM VASCULAR SMOOTH MUSCLE CELLS

PDGF from the two sources discussed above (the platelet and the vascular endothelial cell) may possibly function in vivo as a paracrine hormone—stimulating nearby responsive cells, including vascular smooth muscle cells. We have recently found that under certain circumstances smooth muscle cells may themselves synthesize PDGF-c [24]. In these cases, PDGF-c could be acting as an "autocrine" hormone [8]—stimulating the smooth muscle cells that produce it.

Smooth muscle cells cultured from the tunica media of normal adults of all species examined to date (rat, human, monkey, cow, rabbit; reviewed in [5]) express PDGF receptors and do not secrete amounts of PDGF-c that can be readily detected without concentration of the conditioned medium. However, SMC cultured from the aortae of 2-wk old ("pup") rats display many fewer PDGF receptors per cell and secrete mitogenic levels of PDGF-c [24]. The rate of accumulation of PDGF-c in pup SMC culture medium is, on the average, 150-fold greater than in adult SMC culture medium. Rat SMC PDGF-c is completely neutralized by antiserum prepared against purified human PDGF. Neutralization of rat SMC PDGF-c requires higher concentrations of antibody than is needed to neutralize human PDGF, but this may only reflect species differences in antigenicity, since neutralization of PDGF in rat whole blood serum requires comparably high concentrations of antibody.

Culture medium conditioned by pup rat SMC is mitogenic for mouse 3T3 test cells. Antiserum against human PDGF completely neutralizes the stimulatory capacity of this conditioned medium. This suggests that the only mitogen secreted by pup SMC, at least when assayed on 3T3 cells, is PDGF-c. Pup SMC-conditioned medium has been assayed for the presence of EGF-like factors and found to contain none [24]. The pup SMC appear to secrete only one mitogenically active molecule and, therefore, differ from vascular endothelial cells, which secrete other mitogens in addition to PDGF-c.

In order for the pup cells to be able to respond to the PDGF-c that they themselves produce, they must also synthesize PDGF receptors. However, ^{125}I-PDGF binding to pup and adult rat SMC revealed that pup rat SMC express far fewer, if any, detectable receptors for PDGF [24]. Exposure of cells that possess receptors for PDGF to exogenous PDGF results in a decrease in cell surface receptors, presumably as a result of the processes of receptor occupation, internalization, and degradation [reviewed in 5]. Since pup rat SMC synthesize PDGF-c, it is possible that they also synthesize the PDGF receptor at the same rate as do adult rat SMC, but that these receptors are constantly being occupied and internalized, resulting in fewer detectable receptors and a constant mitogenic stimulation.

The difference in age between the donors of the pup (2 wk) cells and adult (3 mo) cells is only 2½ mo. During this period in vivo, the aorta increases in size as a result of an increase in both the number of cells it contains and of the amount of extracellular matrix (collagen, proteoglycans, etc) present. By 3 mo, rates of both cell proliferation and production of extracellular matrix have essentially ceased. Pup and adult SMC have been cultured ex vivo in a state of continuous proliferation for at least 6 mo without any evidence that the pup cells are acquiring an adult phenotype or vice versa. This suggests that some signal occurs in vivo, but not in culture, that is responsible for this phenotypic conversion. This signal could be a hormonal factor, or could be a response of the SMC to gradually changing character of their local environment—a change in which they themselves could play a major role via synthesis and deposition of extracellular matrix components.

ONCOGENIC TRANSFORMATION

A second circumstance in which autocrine production of PDGF-c may play a role in the proliferative state of cells is oncogenic transformation. We have found that production of PDGF-c can accompany transformation of cells by a wide variety of transforming agents, including the retroviruses Kirsten and Moloney murine sarcoma virus, and simian sarcoma virus; the DNA viruses SV-40 and adenovirus; and a spontaneous transformation event [19]. Production of PDGF-c thus contrasts with production of EGF-like molecules, where production seems to be restricted to cells transformed by certain retroviruses.

Like PDGF-c produced by normal cells, the PDGF-c produced by transformed cells is a true binding competitor and not just an agent that damages PDGF receptors [19,22]. It differs from PDGF-c from normal cell types in that only a portion of the PDGF-c activity is neutralized by the antibody to human PDGF; the remaining PDGF-c is poorly recognized, if at all [19]. It is possible that the inability of anti-PDGF antiserum to recognize some of the PDGF-c produced by transformed cells reflects post-transcriptional modifications of the product synthesized from the PDGF gene. These modifications may be sufficient to alter antigenicity without affecting ability to bind to the PDGF receptor. Alternately, the PDGF-c not recognized by the antiserum may be the product of a different gene and only distantly related to PDGF in amino acid sequence. In this latter case, the PDGF-c from transformed cells would be analogous to the sarcoma growth factor, which is able to bind to the EGF receptor but is the product of a separate gene and is antigenically distinct from EGF [25].

A recent observation, which may shed some light on the nature of the PDGF-c produced by one type of transformed cell, is that the transforming gene product of simian sarcoma virus is nearly identical in amino acid sequence to one of the two peptide chains that constitute PDGF [26,27]. This is the first known instance of an oncogene encoding a growth factor. A reasonable inference would be that the mechanism of transformation by simian sarcoma virus involves autocrine production of the virally encoded PDGF-c [28]. It must be remembered, however, that PDGF-c is produced by many cells transformed by oncogenic agents that do *not* themselves encode a PDGF-like molecule [19]. In these cases, the source of the PDGF-c must be a cellular gene the expression of which is activated by the transformation. It is possible that expression of PDGF is a key event in transformation, and that different transforming agents accomplish this in different fashions: by introducing an anologous gene of their own, or by activating the cellular homolog.

CONCLUSIONS

The data discussed above suggest that PDGF, or PDGF-like molecules using the PDGF receptor, may play roles in several forms of growth regulation. In all systems investigated to date, it appears that PDGF (and PDGF-c) function as locally active factors—either paracrine or autocrine. This might be necessitated by the broad distribution of PDGF-responsive connective tissue cell types, such that control over the sites of PDGF action is exercised by restricting the exposure of tissues to PDGF, rather than by restricting the distribution of PDGF receptors.

REFERENCES

1. Balk SD: Proc Natl Acad Sci USA 68:271, 1971.
2. Ross R, Glomset J, Kariya B, Harker L: Proc Natl Acad Sci USA 71:1207, 1974.
3. Kohler N, Lipton A: Exp Cell Res 87:297, 1974.
4. Westermark B, Wasteson A: Exp Cell Res 98:170, 1976.
5. Bowen-Pope DF, Seifert RA, Ross R: In Boynton AL, Leffert HL (eds): "Cell Proliferation.' New York: Academic Press, (in press).
6. Ross R, Glomset JA: New Engl J Med 295:369, 420, 1976.
7. Dockray GJ: Fed Proc 38:2295, 1979.
8. Sporn MB, Todaro GJ: New Engl J Med 303:878, 1980.
9. Bowen-Pope DF, Ross R: Biochem Biophys Res Com 114:1036, 1983.
10. Oka Y, Orth DN: J Clin Invest 72:249, 1983.
11. Childs CB, Proper JA, Tucker RF, Moses HL: Proc Natl Acad Sci USA 79:5312, 1982.
12. Assoian RK, Komoriya A, Meyers CA, Miller DM, Sporn MB: J Biol Chem 258:7155, 1983.
13. Heldin CH, Westermark B, Wasteson A: Exp Cell Res 136:255, 1981.
14. Adamson ED, Rees AR: Mol Cell Biochem 34:129, 1981.
15. Bowen-Pope DF: In Broder L (ed): "Developmental Biology." New York: Plenum Press, (in press).
16. Roberts AB, Anzano MA, Lamb LC, Smith JM, Sporn MB: Proc Natl Acad Sci USA 78:5339, 1981.
17. Anzano MA, Roberts AB, Meyers CA, Komoriya A, Lamb LC, Smith JM, Sporn MB: Cancer Res 42:4776, 1982.
18. Assoian RK, Grotendorst GR, Sporn MB: Fed Proc 43:802, Abstr 3024, 1984.
19. Bowen-Pope DF, Vogel A, Ross R: Proc Natl Acad Sci USA 81:2396, 1984.
20. Gajdusek C, DiCorleto P, Ross R, Schwartz SM: J Cell Biol 85:467, 1980.
21. DiCorleto PE, Gajdusek CM, Schwarts SM, Ross R: J Cell Physiol 114:339, 1983.
22. DiCorleto PE, Bowen-Pope DF: Proc Natl Acad Sci USA 80:1919, 1983.
23. DiCorleto PE: Exp Cell Res 153:167, 1984.
24. Seifert RS, Schwartz SM, Bowen-Pope DF: Nature 311:669, 1984.
25. Marquardt H, Hunkapiller MW, Hood LE, Twardzik DR, DeLarco JE, Stephenson JR, Todaro GF: Proc Natl Acad Sci USA 80:4684, 1983.
26. Doolittle RF, Hunkapiller MW, Hood LE, Sevare SG, Robbins KC, Aaronson SA, Antoniades HN: Science 221:275, 1983.
27. Waterfield MD, Scrace GT, Whittle N, Stroobant P, Johnsson A, Wasteson A, Westermark B, Heldin CH, Huang JS, Deuel TF: Nature 304:35, 1983.
28. Deuel TF, Huang JS, Huang SS, Stroobant P, Waterfield MD: Science 221:1348, 1983.

Neurobiology: Molecular Biological Approaches to Understanding Neuronal Function and Development 151–168 (1985)

Dynamic Synaptic Interactions During the Formation of a Retinotopic Map

Martha Constantine-Paton and Thomas A. Reh

Department of Biology, Princeton University, Princeton, New Jersey 08544

Recent studies in our laboratory using the developing visual projection of the leopard frog Rana pipiens have shown that axon terminals from the ganglion cells of the retina continuously change the tectal neurons with which they are in functional synaptic contact and that the segregation of two retinas into eye-specific stripes within this tectal lobe is dependent on action-potential activity. Our results are discussed in terms of a model in which a minimum of two independent sets of interactions underlie the establishment and maintenance of the topographic projections between the frog's eyes and its tectal lobes. One involves some form of positional matching between regions of the retina and the developing midbrain. This aligns the normal projection. However, position matching must utilize gradients, rather than absolute pre- to postsynaptic interactions; otherwise synaptic associations could never change throughout development. A second, activity-dependent mechanism probably serves, during normal development, to maintain the relative ordering of moving axon terminals, the cell bodies of which are neighbors in the retina. When two retinas coinnervate one tectal lobe, the activity-dependent mechanism tends to maximize the number of contacts that are made adjacent to contacts from a neighboring ganglion cell, while the positional matching mechanism tends to spread each retinal projection throughout the doubly innervated tectal lobe. The resulting periodically alternating stripes represent the best possible solution to the two mechanistic requirements. The wide distribution of striped termination zones in the vertebrate brain suggests that similar mechanism are common to all topographic projections.

Key words: retinotectal, ocular dominance, TTX activity, sprouting factors

The formation of synapses between afferent axons and their target neurons is often thought of as the last step in the development of neuronal connections. There is growing evidence, however, that the first functional synapses are merely the initiators of the cellular interactions that produce the particular neuronal contacts of the adult brain. In a variety of developing systems, for example, anatomical and physiological observations indicate that many young synapses are transient structures. The distribution of axonal terminal arbors and the number of different inputs impinging on a given cell often become reduced as the animal matures, resulting in the refinement of initially more diffuse projections. It has also been suggested that the widespread occurrence of ontogenetic neuronal death reflects the elimination of cells the contacts of which have been deprived of postsynaptic space during this refinement process.

Received July 23, 1984.

In the retinotectal system of cold-blooded vertebrates, there is little evidence of cell death as a result of synaptic competition. Rather, it now seems clear that the phenomenon of synapse refinement produces a qualitatively different result in this pathway. The processes of axon elongation, synapse formation, and synapse elimination occur concurrently during the development of the retinotectal pathway, and they are organized topographically. Consequently, the termination sites of the entire afferent cell population systematically shift across the field of postsynaptic dendrites.

The suggestion that such a shift must occur during the development of the retinotectal projection arose initially from observations on the normal development of the system in Xenopus. The Xenopus retina grows as an expanding circle from a proliferative zone surrounding its perimeter. The optic tectum, the retina's major target in these animals, grows only at its posterior and medial boundary. Nevertheless, during the period when these structures are growing, the axon terminals of the retinal ganglion cells form an ordered array on the tectal surface [1]. Straznicky and Gaze, who conducted these investigations, reasoned that such topographic order could only be maintained during discordant growth if the projection sites of the earlier-formed ganglion cells were continually displaced toward the posterior and medial border of the optic tectum [2].

Since it was first proposed, the hypothesis of shifting connections has generated considerable controversy, both because the initial studies provided no direct evidence for dynamic synapses and, moreover, because the concept of chemospecificity—that source and target neurons at corresponding positions have particular matching molecular labels [3]—was firmly entrenched [4]. In addition, subsequent, more detailed studies of retinal growth in Xenopus have demonstrated that a marked asymmetry in the addition of new cells to the retina brings its growth more into line with that of the tectum. While several more direct tests of the shifting connections hypothesis have been attempted in Xenopus, none have been conclusive [5–7]. Recently, Fraser demonstrated that the site of light-evoked action potential activity recorded in the tectum following stimulation of central retinal cells shifts to a more caudal tectal position as the animal matures; however, the shifts in termination sites that occur are relatively small. Shifts in the position of a projection on the order of 100 microns or less were reported, leaving open the possibility that with dendritic elaboration and a modest amount of trimming of the presynaptic arbor an apparent shift could occur without any real change in pre- and postsynaptic partners [8,9]. In addition, a number of laboratories working on both fish [10,11] and amphibians [12] have analyzed the trajectories of retinal ganglion cell axons as they grow through and form terminal arbors within the tectal neuropil. The morphology and orientation of these axon terminals essentially represent a structural record of where the ganglion cell processes first entered the tectum and where they initially arborized around tectal dendrites in relation to where their final arbor resides. These studies are consistent with the notion that the terminal arbor zone shifts posteromedially, relative to the tectal surface. Nevertheless, studies of the afferent axons alone cannot reveal whether an actual shift of pre- and postsynaptic partners is occurring in these projections. The tectum changes gradually in shape, with continuing growth making it difficult to determine with any certainty that the ganglion cell termination sites shift while the relevant postsynaptic tectal dendrites remain behind.

In this overview, we will describe the evidence from our laboratory that ganglion cell axon terminals do shift in their projection position relative to tectal neurons during larval development, thus indicating the dynamic nature of this projection in the frog tadpole. Within this context, we will propose a model to explain the ability of retinal ganglion cell terminals to maintain their relative positions in this highly ordered projection during the substantial shifts in their tectal locations, and we will present some recent evidence that supports this model. We will also compare the results obtained in this system with those from studies of synaptogenesis in the neuromuscular junction and try to gain some insight into the possible molecular mechanisms underlying topographic central nervous system connections.

MATERIALS AND METHODS

Animals

All animals were lab-reared Rana pipiens, obtained by ovulation and artificial fertilization using captured adult wild Rana pipiens obtained from suppliers (Hazen and Sons, Alburg, VT; West Jersey Biological Supply, Wenonah, NJ). Embryological microsurgical procedures were employed to produce the three-eyed frog [13].

Anatomy and Physiology

All procedures have been detailed in other publications [12,14,15]. We have used ^{3}H-thymidine (^{3}H-Tdr) to birthdate retinal and tectal cells, axonal transport of ^{3}H-proline or horseradish peroxidase (HRP) (followed by autoradiographic processing or histochemical localization in tissue sections or whole-mounted brains) to determine the position of ganglion cell terminals in the tectum, and extracellular physiological recordings from perfused tadpoles to examine functional aspects of this system during development.

Pharmacology

Tetrodotoxin (TTX) was obtained from Sigma (St. Louis, MO) and infiltrated into ELVAX plastic (DuPont, Wilmington, DE). Its rate of release from ELVAX pellets was roughly calibrated, using equal concentrations of fluorescein, which has an approximately equal molecular weight (329 fluorescein vs 313 TTX). TTX-infiltrated ELVAX was either implanted as an intraocular pellet or tied around the optic nerve as a cuff. Effectiveness of the TTX block of retinal ganglion cell activity was ascertained by physiological recordings from the tectum or by monitoring the animal's optokinetic response. Kainic acid was obtained from Sigma and injected as a 10^{-3} M solution into the vitreous of the eye (1 μl). Effects of kainic acid on the retina were monitored at least 2 wk later by cell counts in each of the retinal layers and by physiological recording from ganglion cells in the retina and optic tract.

RESULTS

Shifting Retinotectal Projections

The first retinotectal ganglion cell axons reach the anterolateral border of the midbrain of frog larvae roughly one day after neurulation. At this age, the lens of the eye has not yet cleared, and few, if any, postsynaptic neurons are present in the primordia of the optic tecta [16]. Injections of ^{3}H-Tdr at very early neural tube stages of development are found, upon examination at later stages, to label ganglion cells in the dorsal-posterior quadrant of each retina, immediately adjacent to the optic nerve head, where all ganglion cell axons leave the eye. However, a ^{3}H-Tdr pulse administered at any stage after the embryo becomes a free-swimming tadpole labels an annulus of cells that is fairly concentrically positioned around the optic nerve head. Moreover, when a tadpole is subjected to successive staggered injections of ^{3}H-Tdr during tadpole life, it becomes clear that the rate of cell division is symmetrically distributed around the perimeter of the retinal epithelium at any particular tadpole stage, and that there is essentially no cell death in the ganglion cell layer during the larval period [14].

Similar ^{3}H-Tdr studies of the optic tectal lobes reveal the first postmitotic tectal neurons at the anterior lateral edge of the midbrain, coincident with the arrival of the first retinal ganglion cell axons [16]. Thereafter, postmitotic cells are laid down as a crescent-shaped band, stretching along the posterior medial border of the differentiating tectal epithelium. Tectal neurogenesis is essentially complete in the young postmetamorphic frog [17]. The patterns of retinal and tectal growth in Rana pipiens tadpoles are summarized in Figure 1.

Using electrophysiological recording techniques and neuroanatomical tracing of small subpopulations of retinal axons, we have found that all the visual response classes of retinal ganglion cells documented in the adult are present in the retinas of young, free-swimming tadpoles. Moreover, chemically mediated excitatory postsynaptic potentials and a high-fidelity, appropriately oriented representation of the retinal surface is produced by the axons of these ganglion cells within the anterior, differentiated, regions of the tectal lobe [15]. In short, a visual pathway between the center of the retina and the anterior fraction of the tectal lobe is well established in these tadpoles at a period when most of the peripheral retina and the more posterior tectal neurons are not yet postmitotic. This pathway already possesses the major functional characteristics and the finely tuned topographic map of the mature frog.

A previous study in Rana had detailed the organization and position of axons from different regions of the retina in the visual pathway [18]. Consequently, it was possible to surgically expose the optic tract in the diencephalon and reliably disrupt only the small bundle of axons arising from the oldest ganglion cells distributed around the optic nerve head. Selective labeling of these axons was accomplished by placing a small pellet of HRP at the site of the disruption. After a 24-hr survival period, the retina was flattened, the brain was dissected out of the skull, and both pieces of tissue were processed without sectioning to visualize the transported HRP. Horseradish peroxidase travels rapidly and efficiently in both anterograde and retrograde directions from the site where an axon is cut; consequently, in this material we were able to verify the central position of labeled retinal ganglion cell bodies at the same time that we visualized the positon of their synaptic terminal arbors relative to the anterior tectal borders in the intact brain [15]. The results of measurements on these brains are shown in Figure 2. It can be seen that, during the latter three-quarters of tadpole life (for which data is shown), the projection of central retinal terminals is displaced over a millimeter along the anterior posterior tectal axis. The long dimension of individual axonal terminal arbors in this population is approximately 150 microns; consequently, the movement represents a shift that is over five times the extent of the ganglion cell arbors.

These displacements reflect actual movements of retinal arbors relative to tectal neurons. We demonstrated this by giving tadpoles single injections of ^{3}H-Tdr at early hindlimb bud stage IV. The isotope labeled the crescent-shaped band of tectal neurons

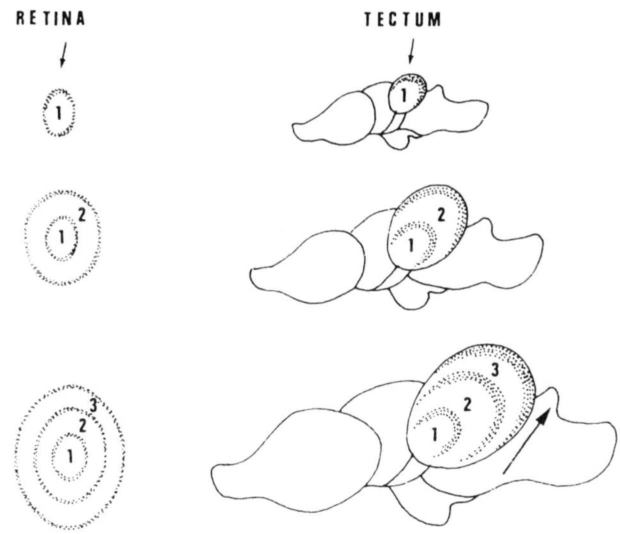

Fig. 1. Noncongruent patterns of cell addition to the frog retina and optic tectum. Stippling illustrates the position of ^{3}H-Tdr-labeled cell somata after three pulses (1–3) of the isotope at successively older tadpole stages. The retina adds cells as rings around its margin, while new cells are added to the tectum in a crescent on its posterior and medial border.

that became postmitotic at that stage. The HRP procedure that documented the projection site of the central retinal ganglion cells was applied to groups of these animals before killing them at successively later tadpole stages. The brains were subsequently sectioned, processed for light microscopic autoradiography, and both the band of ^{3}H-Tdr-labeled tectal neurons and the projection site of central retinal ganglion cells were mapped on a scaled reproduction of the tectal surface, using a computer graphics system. In early and mid-larval stage tadpoles, central retinal ganglion cells project to sites in front of the band of labeled neurons. However, in early postmetamorphic frogs, the oldest ganglion cells inevitably project to positions located behind the labeled band of tectal cells. The results of this experiment are diagrammed in Figure 3.

The dynamic nature of this tectal synaptic zone is reflected in its cellular morphology. Electron micrographs reveal growth cones on processes that are also clearly the presynaptic components of synapses (Fig. 4). However, these same electron micrographs provide little evidence of presynaptic release sites without opposed postsynaptic profiles. In addition, camera lucida reconstructions of individual ganglion cell terminals show growth-cone-like structures, and these filopodia are concentrated on the posterior edges of the terminals' arbors [15]. Such observations suggest that presynaptic specializations are not simply dragged through the retinotectal neuropil; rather, the predominate mode of terminal movement is a creeping process, where synapses and their supporting branches retract from the more anterior regions of the terminal, while new branches are constantly sprouting from the posterior portions of the terminal arbor.

These results indicate that the process of achieving specific neuronal connections is only partly the result of the type of developmental determination usually envisioned

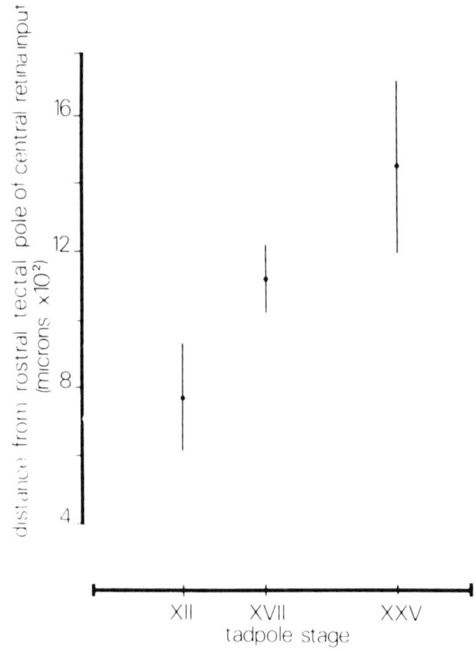

Fig. 2. Shifts in the tectal projection sites of the central and oldest retinal ganglion cells were measured in whole brains after selectively labeling axons from a central retina with HRP. Each point represents the results of measurements in six animals, and each measurement was made by extending a small fiber from the anterior pole of the tectum to the middle of the HRP-labeled terminals, then measuring the length of the fiber with a planimeter. The stages are those of Taylor and Kollros (1964) and refer to the morphological development of the hindlimb. The total duration of the period during which measurements were taken was about 3 mo.

as a position-dependent expression of particular adhesive molecules on both the source and target neurons. While a detailed discussion of the evidence for regionally varying molecular "labels" or "markers" in this system is beyond the scope of this review, a few general points are in order. Several studies in frogs and fish of both regenerating [19] and initially developing optic fibers [12,20] show that axons from a given region of retina will grow into and innervate the appropriate area of the tectum, regardless of the route the fibers take to get from the eye to the tectal lobes, and independently of the presence of any Na^+-dependent action-potential activity in the nerve [21]. In vitro studies, using organized co-cultures of chick retina and tectum, have shown a similar position-dependent selectivity between ganglion cell growth cones and their appropriate tectal cells [22]. Thus, while there is considerable evidence that retinal cell axons and tectal cells from corresponding regions do show a selective affinity for one another, it is clear that a mechanism employing rigid lock-and-key type adhesive interactions between cells from corresponding regions of retina and tectum cannot work in a system where connections are continually shifting. Instead, it is likely that the molecule(s) that mediate the selective cellular adhesions observed in this system are distributed in a graded fashion. The notion that the "labels" for retinotectal specificity are graded, rather than fixed, was originally proposed by Gaze and his coworkers to explain the results obtained when "compound" eyes were formed from

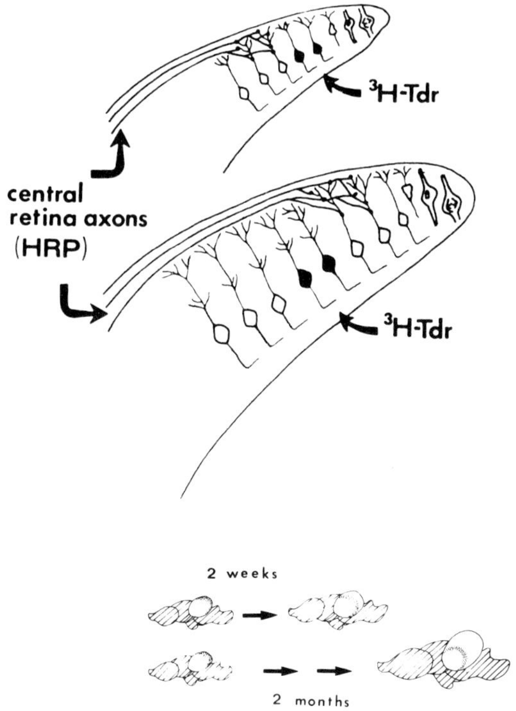

Fig. 3. The shift of oldest retinal ganglion cell terminals from a position in front of a contiguous band of labeled tectal cell somata to a position behind the equivalent band of tectal cell somata when examined in an older tadpole. The inset illustrates the procedure for pulsing all animals with ^{3}H-thymidine (^{3}H-Tdr) at one developmental age and killing them at successively older ages, so that the ^{3}H-Tdr constitutes a stable label identifying the cells that become postmitotic when the ^{3}H-Tdr was initially given. In the figure the tecta of two different animals, both pulsed with ^{3}H-Tdr at the same early tadpole stage, are illustrated in a diagrammatic cross-section parallel to the anterior-posterior axis of the lobes. The upper section illustrates the tectum of an animal killed 2 wk after the ^{3}H-Tdr pulse; the lower section illustrates the tectum of an animal killed 2 mo after the pulse. Tectal neurons at various stages of differentiation are illustrated on the posterior half of both cross-sections.

identical halves of donor and host optic anlagen [23]. Graded labels have subsequently been successfully incorporated into several computer models of retinotectal map formation [24–26]. At this point, the molecule or molecules involved in the selective adhesion exhibited between corresponding retinal and tectal cells have not been identified, although a recent report of monclonal antibodies with a topographically graded distribution of binding demonstrates that these explanations derived from cell behavior might have a physical basis [27].

The Function of Action-Potential Activity

However, a system of topographically graded adhesive molecules cannot explain all of the results obtained from experiments with the retinotectal system. We discovered a number of years ago that when two eyes are forced to innervate a single tectal lobe, either by the embryonic transplant of an additional eye primordium or by the

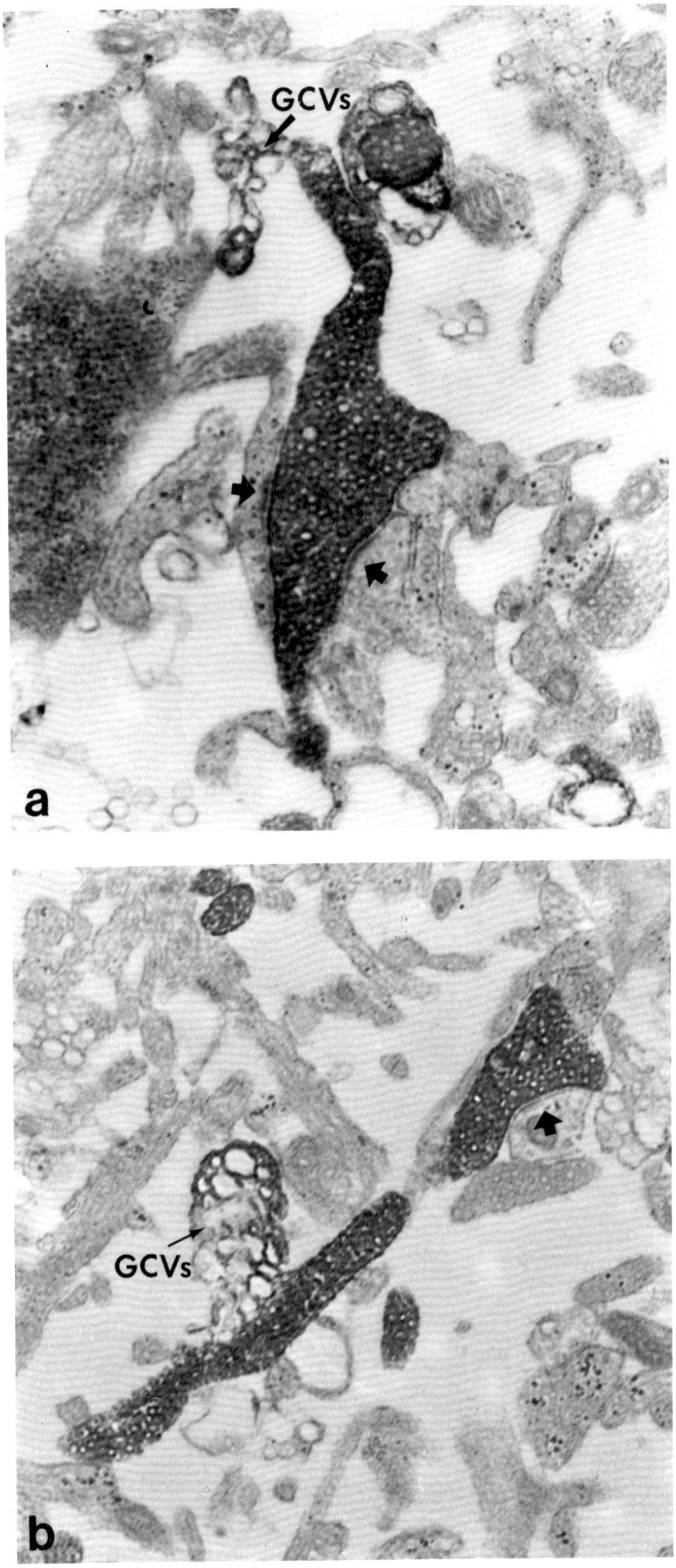

Fig. 4. Electron micrographs of optic nerve processes, anterogradely filled with HRP. These processes share the characteristics of growth cones (growth cone vesicles (GCV)) and of synaptic contacts with tectal cell dendrites (arrows). Original magnification: a, ×37,000; b, ×32,000.

surgical removal of one tectum in a normal frog or tadpole, the terminals from the two eyes divide the tectal neuropil into a series of alternating, mutually exclusive, eye-specific bands [13,14]. This experimentally induced pattern is reminiscent of the ocular dominance columns of mammalian cortex. However, because the frog normally never exhibits this segregation, it seemed likely that the mechanisms involved were the same ones active in forming the single continuous retinotopic map [28].

In the model that links the segregation of afferents into eye-specific strips to the normal process of topographic map formation, the mapping process involves two mechanisms. A graded distribution of relative adhesiveness along both the rostrocaudal and mediolateral dimensions serves to align the projection and give it a roughly correct topography, while a competitive process, which stabilizes adjacent terminals from neighboring retinal ganglion cell bodies at the expense of terminals from non-neighboring cells, acts to "fine-tune" the retinotectal projection to its characteristic high degree of retinotopy [28]. Similar models, incorporating two or more mechanisms, have also been advanced by others, and computer simulations of these hypotheses can predict the behavior of this system under experimental perturbations [29–31]. This two-stage mapping mechanism will form stripes when two separate topographic projections are forced to terminate in the same target. The position-dependent adhesions will tend to align both projections and spread them across the entire target zone, while the competitive mechanism will cause segregation into stripes. Only a striped pattern maximizes the contacts each retina can have within the appropriate region of the target, while also minimizing the number of times a ganglion cell synapse is positioned adjacent to a synapse from a "non-neighboring" ganglion cell from the other retina. The question that arises immediately from this explanation is what type of mechanism underlies the competitive phase of map formation? Several possible origins for this eye-specific information have been examined in prior publications. Intrinsic differences in laterality have been shown to be uninvolved, since segregation will occur between two eyes of the same original laterality in three-eyed frogs [28,32]. Histocompatibility differences between donor and host have been ruled out by the fact that the two eyes of a normal frog [14] or goldfish [33] can be made to segregate when forced to innervate a single tectum. Selective fasciculation of fibers in the optic tract is also unlikely to provide source-specific information, since the terminals that comprise one band have been shown to travel to the tectum in numerous different age-related bundles [12]. Furthermore, experiments using compound eyes formed by fusing embryonic retinal fragments [34–36] or by inducing mirror-symmetric duplication of eye primordia by partial ablation [37] have demonstrated that ganglion cell arbors from the two half-eyes will segregate in the tectum. This indicates that axon terminals from cells of a single continuous presynaptic sheet will segregate in the tectal neuropil, depending solely on differences in the position of the ganglion cell somas.

Evidence from studies of cat and monkey visual cortex suggests another possibility: that the action potential activity patterns in the optic nerve might provide the basis for segregation. Experimental manipulations that affect the level of activity in the retinal ganglion cells (such as monocular lid suture or dark rearing) also affect the ocular dominance columns in the visual cortex. Therefore, we reasoned that activity might also play a role in the experimentally induced segregation of retinal ganglion cell terminals in dually innervated tecta. To test this hypothesis, three-eyed Rana pipiens tadpoles were raised in a dark box from Shumway embryonic stages to mid-larval tadpole stages. There was no noticeable change in the striped pattern formed by the two eyes, despite the complete absence of pattern vision—or even any light—during their development.

However, this experiment does not rule out the involvement of retinal ganglion cell activity in stripe formation, since even in the dark these cells are spontaneously active. Moreover, there is considerable evidence in several species that the spontaneous activity patterns of ganglion cells of the same response type are highly correlated when their receptive fields overlap but are not correlated between disparately positioned cells [38–43]. Figure 5 presents some recent data from Rana pipiens tadpoles indicating that ganglion cells of similar response type and with overlapping receptive fields have temporally correlated patterns of activity in this species as well. Simultaneous recordings of the spontaneous activity from two "changing contrast detector" [44] retinal ganglion cells with substantially overlapping receptive fields are shown in Figure 5a and b. Figure 5c presents one way of illustrating the correlation in activity between these two units. The figure contains a post-event histogram generated by using one spike of the first cell as a trigger for the collection of succeeding action potentials from the second neuron.

It is likely, therefore, that the proximity of retinal ganglion cell bodies is consistently related to the degree of correlation in the timing of their activity. Hebb, in 1949 [45], was the first to suggest that neural activity might serve to selectively stabilize a particular input to a cell if the input's firing was correlated with the firing of the cell. The temporal summation provided by the simultaneous activation of two (or more) presynaptic fibers with synapses on the same cell would increase the likelihood of postsynaptic activation and concomitant stabilization of both inputs [46,47]. Alternately, if the two presynaptic fibers have asynchronous patterns of activity, their connections with the same cell would not be strengthened. Retinal proximity-dependent sorting of synapses could therefore occur if the postsynaptic cell (or circuitry) could somehow reinforce these effective synapses at the expense of contacts from non-neighboring ganglion cells that might fire with equal frequency but in an uncorrelated pattern [18,30,31].

In the developing tadpole tectum, the neighborhood similarities in the pattern of action potential activity across the retina could provide the kind of continuously acting cue necessary to maintain the high degree of internal order in the projection as individual pre- and postsynaptic relationships changed. In other words, in a situation where all retinotectal synapses are constantly turning over, selection would be for terminals that converge on dendrites along with terminals from their nearest retinal neighbors. This would maintain the coherence of the projection, because the number, the surface area, or the lifetime of new cooperative contacts would be consistently increased over new contacts from relatively unsynchronized and therefore less cooperative inputs. This line of reasoning led us to predict that the pattern of stripes that form upon double innervation of tadpole tecta would probably also change with time, because they are composed of constantly changing contacts. Moreover, each stripe probably represents a separate zone of functional cooperation among groups of neighboring ganglion cells in each retina. We addressed this possibility by applying TTX to the retinas of tadpoles in which double innervation of a tectal lobe had been induced by implanting a third eye primordia into early embryos. Earlier studies of three-eyed tadpole preparations had revealed that eye-specific segregation is present in the tectal lobes of these animals from the earliest tadpole stages. However, if

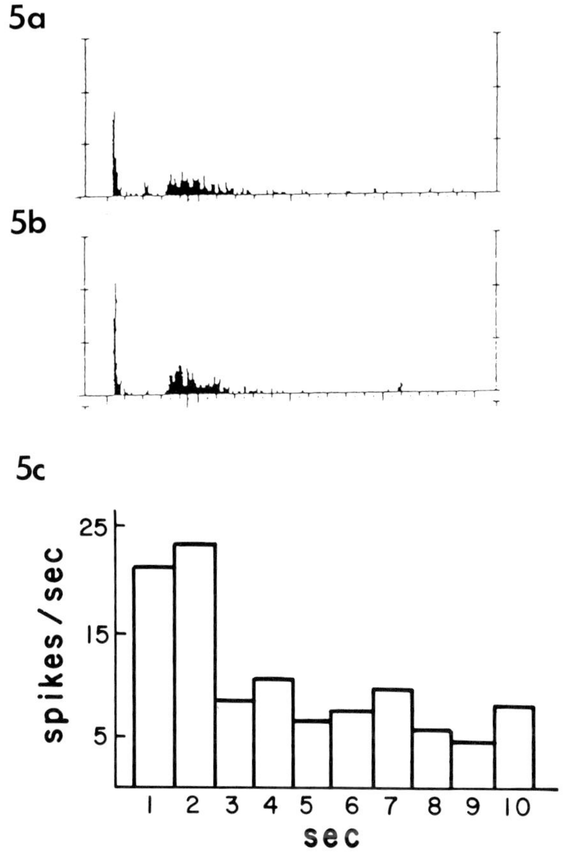

Fig. 5. Temporal correlations in action-potential activity between two changing contrast activated retinal ganglion cells in the eye cup of a frog. The two cells were separately isolated and shown to have overlap of approximately 50% of their receptive fields. Their post-stimulus time histograms (a,b) are shown for 20 stimulus (40 μsec light flash) presentations. The recording room was subsequently completely darkened, and the output of the spontaneous activity of each of the two units was collected for 20 triggering events by using an action potential of the upper cell (a) to trigger the collection of succeeding action potentials in the second neuron (b). This produced the one-sided temporal correlation of the activity in the two units shown in c, and it is observed that there is a greater likelihood that the units will fire within 2 sec of one another than chance would predict.

ganglion cell action potential activity provides the cues used at the terminals of these cells to selectively favor some new contacts over others, prolonged silencing of the retinas of these three-eyed frogs (without any disruption of the anatomical pathway) should cause a progressive desegregation of their doubly innervated tectal lobes. The results of this experiment are shown in Figure 6. The striped projection zone of the third eye has been labeled with HRP in the upper photograph. The lower photograph also illustrates the tectal projection of a supernumerary eye in a similarly staged tadpole. However, there is only a vestige of stripe formation in the latter tectal lobe. Both of these animals also received intraocular injections of ^{3}H-proline in their normal competing retinas. Subsequent histological processing and light microscopic autora-

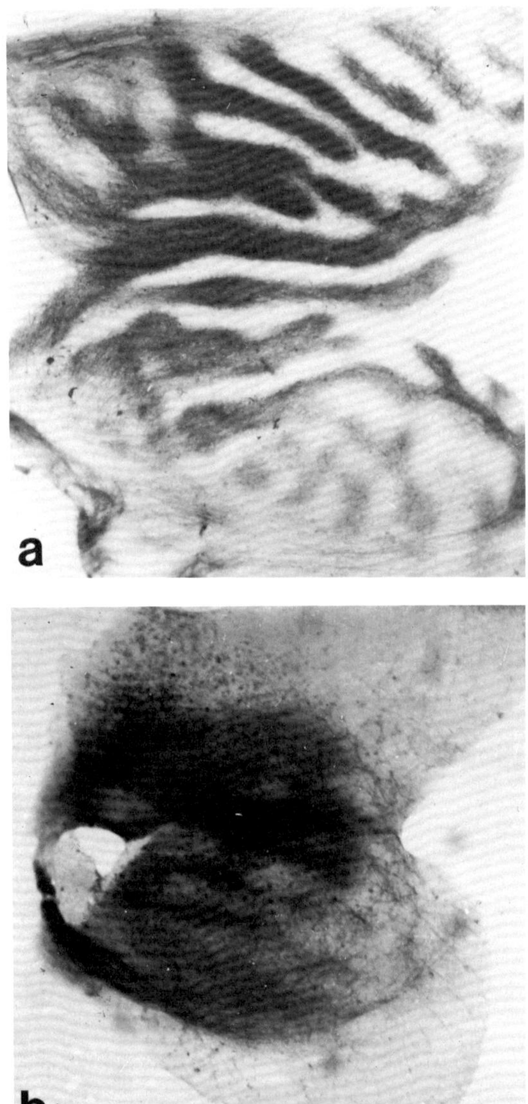

Fig. 6. Flat-mount preparations of optic tecta from two three-eyed tadpoles following anterograde labeling of the third optic nerve with HRP. In a the tadpole had all three optic nerves silenced with TTX for 1 wk. While there is some evidence of debanding in the rostral tectum, most of the tectum shows the distinct striped pattern observed in untreated three-eyed tadpoles. In b the tadpole had all three optic nerves silenced with TTX for 3 wk. There is now almost complete desegregation, with only a slight periodic variation remaining. Rostral is to the left in both a and b.

diography revealed ^{3}H-proline-labeled bands interdigitating with the HRP-labeled bands in the upper tectal lobe. However, the autoradiographs of the lower tectal lobe demonstrate a dense projection from the contralateral host eye that completely overlapped the continuous HRP-labeled projection from the supernumerary eye. The animal whose brain is shown in Figure 6a survived for 14 wk with cuffs of ELVAX, a slow release plastic, on all three of its optic nerves. The brain in Figure 6b came from an animal treated in exactly the same manner, except that the ELVAX cuffs were maintained for 4 weeks. In both cases the ELVAX was infiltrated with concentrations of TTX sufficent to release 10^{-5}–10^{-7}M quantities of the drug for over 3 wk. Behavioral tests, electrophysiological studies on two-eyed animals, and histological examination of retinas (including ganglion cell counts) verified that these quantities of TTX are effective blockers of all retinal ganglion cell activity, yet they had no morphological effect on the retinas [48].

Evidence of the process underlying this breakdown in eye-specific segregation was obtained when we cleared, flattened, and microscopically examined the initially striped tectal lobes of tadpoles subjected to varied periods of survival with the TTX block. As early as 2.5 wk after the onset of the block, numerous fine branches from the terminals of ganglion cells in the HRP-labeled supernumerary projection could be observed in the interband zone (Fig. 7a), and this is in marked contrast to the interband region of striped tecta with normally active retinal inputs (Fig. 7b). Camera lucidae reconstructions of these terminals and measurements of their tangential extent reveal that they cover, on average, nearly twice as much tectal area as arbors in active projections. Moreover, the numerous growth-cone-like structures characteristically predominating on the posterior side of the arbor in normal tadpoles and three-eyed animals with active projections are distributed around the entire circumference of the arbors of these silenced retinal ganglion cells [48].

Most recently, we have found that a single 1 μl injection of a 10^{-3} M kainic acid solution into the posterior eye chamber of these tadpoles can be used to destroy amacrine (and probably also some bipolar neurons) in their retinas without any morophological or numerical effect on retinal ganglion cells. The treatment applied to all three-eyed animals appears to mimic the debanding effect of TTX. The significant aspect of this treatment is that it alters but does not eliminate the patterning of action-potential activity transmitted to the tectal lobes. We consistently find a pronounced decrease in the amount of sustained action-potential activity that all types of the frog retinal ganglion cells produce in response to their preferred stimuli. Although details of the altered neuronal responses remain to be explored, neurochemical retinal lesions such as these promise to be powerful tools with which to subtly alter normal retinal function and dissect out exactly what property of the ganglion cells' firing pattern conveys the nearest neighbor information so crucial to the spatial organization of the dynamic synaptic interactions in their tectal targets.

MOLECULAR MODELS FOR ACTIVITY-DEPENDENT SYNAPTIC INTERACTIONS

The activity-dependent aspect of retinotectal map formation has several parallels to the later phase of synapse formation between motoneurons and muscle fibers. Like the retinotectal projection in the neuromuscular system, there is a continual remodeling of motoneuron axon terminals during development. Some synapses are withdrawn, while at the same time new junctions are being formed. This process of

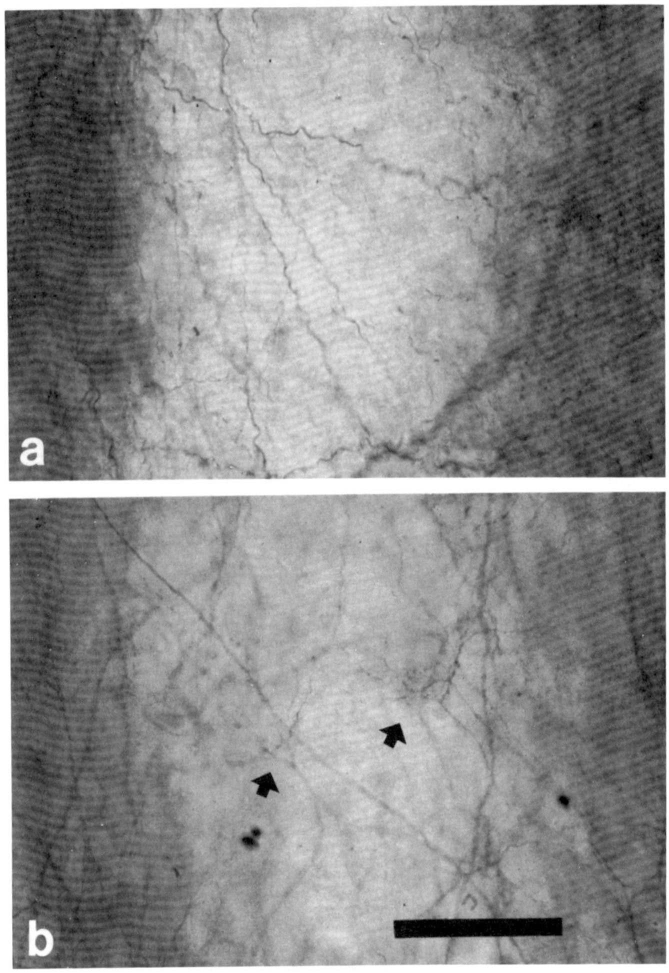

Fig. 7. Sprouting of axonal arborizations into the interband regions of striped tectal lobes after TTX silencing of the projections. a) A photograph illustrating two striped termination zones and an interband region after filling one retinal projection with HRP in a cleared and flattened tectal lobe. Only labeled axons of passage are visible in the interband region in this lobe, with two normally active retinal projections. b) A similar view of a striped tectum in an animal where both innervating optic nerves had been continuously blocked with TTX for 2.5 wk. Arborization (arrows) now can be observed in the interband zone. Scale bar = 50 microns.

remodeling appears to involve an interaction among terminals that ultimately leads to the elimination of some of the inputs to a given muscle fiber, as well as a reduction in the number of muscle fibers innervated by a given axon [49]. This developmental elimination of inputs is slowed by application of drugs such as curare [50], TTX [51], and α-bungarotoxin [52]. All of these prevent activity in the nerve or muscle. Conversely, the elimination of inputs is accelerated by direct electrical stimulation of the nerve or muscle [53]. A definitive explanation for the competitive interactions that remodel the neuromuscular innervation is not yet at hand. However, a number of proposals have been advanced to account for the results of experiments with this system. The two major classes of explanation differ in that the interaction they postulate is mediated by diffusible substances or by actual changes in some membrane or extracellular matrix component. A diffusible trophic factor released into the intercellular space by uninnervated or inactive muscle fibers could explain the fact that nearby intact terminals sprout new collaterals to experimentally denervated fibers. The same diffusible factor could selectively promote the survival of a particular endplate on an innervated muscle fiber if that endplate became more effective in taking up the substance, as a result of its size or its location relative to competing fibers or to its greater amount of activity [54]. Conversely, O'Brien et al [55] have proposed that activity at the neuromuscular junction results in the release of a proteolytic enzyme that digests the nerve terminals on it. This idea requires that while the axon is constantly replacing the terminals, only a limited number of terminals can be maintained by any single axon, due to limits on the transport of critical structural components. As a result, terminals that are replaced more rapidly are favored, while those replaced at a slower rate are eventually completely digested. If a muscle fiber has no innervation, it releases no proteolytic enzyme, and new sprouts from neighboring terminals can innervate it.

The other class of models involves alterations in non-diffusible factors, such as membrane-potential changes, acetylcholine receptor clustering, or extracellular glycoproteins to mediate the competition among fibers. Particularly significant in this respect is an attempt by Stent [56] to provide a molecular basis for Hebb's hypothesis of activity-mediated stabilization of synapses. In this model, when an action potential in the presynaptic cell is effective enough to cause an action potential in the muscle fiber, the drastic shift in membrane potential causes the degradation of synaptic contacts. However, the synapse between the activating presynaptic fiber and the postsynaptic cell is protected from this degradation, since the excitatory postsynaptic potential (EPSP) generated at this synapse results in a much smaller potential change than that from an action potential.

It is clear that these suggestions are still far from a molecular explanation. However, applying them to the formation of retinotectal synapses may prove to be a fruitful approach to elucidating the mechanism of competitive interaction in this system. Furthermore, the models that invoke diffusible factors for activity-dependent synaptic interactions, in particular, can simultaneously account not only for refinement in the frog's retinotectal map and eye-specific segregation but also for the progressive orderly shift of the entire projection across the developing tectal lobe. For example, the trophic substance proposed under the first scheme could serve to draw existing retinal ganglion cell terminals backward toward newly generated tectal cells, setting up an intratectal chemotactic gradient. The concentration of the substance would be high in caudal tectum because it would be released by the new, relatively uninnervated tectal neurons. In addition, the concentration of such a substance would decrease in more rostral regions of tectum if either its release or its removal from the extracellular space is assumed to be proportional to innervation density. If a protease, instead of a trophic factor, is released by tectal cells, its distribution would be exactly the complement of the chemotactic agent. It would be present in high concentration in the active, densely innervated anterior tectum and in relatively lower concentrations in posterior tectum, among the new uninnervated tectal cells. This distribution would tend to digest contacts located along the anterior regions of a retinal ganglion cell's terminal arbor and cause the ganglion cell to eleborate new sprouts along the posterior edge of its terminal field, in order to maintain a relatively stable number of contacts.

It is important to emphasize before closing that these ideas do not replace the need to invoke a two-stage mapping mechanism to work in the retinotectal system. Both the trophic factor and protease hypotheses require strictly position-dependent competition in some form to assure that topographically appropriate retinal ganglion cell arbors create the high innervation density in rostral tectum by continually adding to that already occupied zone.

The trophic factor or protease models do, however, add a set of cell and molecular level constraints for the second activity-dependent phase of map formation. If, for example, there is a limit on the number of postsynaptic sites that can be occupied on a tectal neuron, and the release of the sprouting factor is reduced only when the postsynaptic cell is actually brought to threshold and fires an action potential, then those terminals that are most effective in driving tectal cells will be most effective in limiting the release of the factor into the surrounding medium. This suppression will then prevent sprouting from nearby axonal arbors, and the synapses present on the cells will remain. Coincidently active terminals will be more likely to drive the postsynaptic to threshold (due to temporal summation of their inputs), and therefore will be more effective in suppressing the release of the sprouting factor. In the case of the proteolytic enzyme secretion mechanism, the enzyme will be present to digest terminals only when the cell is brought to threshold, and therefore digestion rates will be higher with coincidently active inputs. The continued release of the protease will require continual replacement of all synapses, but where coincidently active inputs converge to fire the cells, the enzyme will be in highest concentration, and this will decrease the chances that new synapses with uncorrelated activities can invade effectively innervated territory.

Although these models are highly speculative, there is some evidence to support some of the assumptions. For example, it appears from the analysis of synaptic density in three-eyed frogs [57] and in surgically compressed retinotectal projections in goldfish [58] that tectal neurons can effectively limit the number of synapses they receive, despite significant increases in the number of afferents projecting to the tectal region. There is also some reason to believe that coincidently active inputs are favored during neuromuscular junction formation, because in dually innervated muscle fibers the two endplates are more often branches of the same axon than would be expected if the process of elimination were random [59,60]. Finally, the numerous "trophic" substances that have been isolated from end organs [61] are also likely to be localized in central nervous sytem targets. For example, Turner et al [62] have provided evidence that a factor extracted from brain cells increases the outgrowth of neurites from explants of fetal rat retina. More recently, release of proteases from cultured rat ganglion cells has been documented biochemically [63]. Neurobiologists must continue to identify the molecules that are correlated with significant neurobiological events. There is also the additional task of devising assays that can provide conclusive evidence that particular molecules are crucial to the hypothesized mechanisms. However, it is hopefully clear from this review that understanding the interactions among the cells in the system developing in vivo is also essential for appropriately interpreting the role these molecules may play.

ACKNOWLEDGMENTS

Much of this work has been supported by NIH grant EYO1872 and NSF grant BNS8110225.

REFERENCES

1. Gaze RM, Keating MJ, Chung SH: Proc Roy Soc Lond (B) 185:301–330, 1974.
2. Straznicky C, Gaze RM: J Embryol Exp Morphol 28:87–115, 1972.
3. Sperry RW: Proc Natl Acad Sci USA 50:703–710, 1963.
4. Meyer RL, Sperry RQ: In Gottlieb G (ed): "Studies on the Development of Behavior and the Nervous System," Vol. 3. New York: Academic Press, 1976, pp 111–149.
5. Jacobson M: Brain Res 127:55–67, 1977.
6. Scott TM, Lazar G: J Anat 121:485–496, 1976.
7. Gaze RM, Keating MJ, Ostberg A, Chung SH: J Embryol Exp Morphol 53:103–143, 1979.
8. Fraser SE: Dev Biol 95:505–511, 1983.
9. Fraser SE: Soc Neurosci Abstr 9:760, 1983.
10. Rusoff AC: Invest Opthalmol Vis Sci Suppl 18:2, 1980.
11. Easter SS, Stuermer CAO: J Neurosci 4:1052–1063, 1984.
12. Constantine-Paton M, Pitts E, Reh TA: J Comp Neurol 218:297–313, 1983.
13. Constantine-Paton M, Law P: Science 202:639–641, 1978.
14. Law P, Constantine-Paton M: Proc Natl Acad Sci USA 77:2314–2318, 1980.
15. Reh T, Constantine-Paton M: J Neurosci 4:442–457, 1984.
16. Currie J, Cowan WM: Dev Biol 46:103–119, 1975.
17. Currie JR: Thesis, Washington University, St. Louis, 1975.
18. Reh TA, Pitts E, Constantine-Paton J: J Comp Neurol 218:282–296, 1983.
19. Attardi DG, Sperry RW: Exp Neurol 7:46–64, 1964.
20. Fujisawa H, Tani H, Watanabe K, Ibata Y: Dev Biol 90:43–57, 1982.
21. Harris WA: J Neurosci 4:1153–1162, 1984.
22. Bonheoffer F, Huff J: EMBO J 1:427–431, 1982.
23. Gaze RM, Jacobson M, Szekely G: J Physiol 176:409–417, 1965.
24. Prestige MC, Wilshaw DJ: Proc Roy Soc Lond B 190:77–98, 1975.
25. Fraser SE: Dev Biol 79:453–464, 1980.
26. Barbera AJ: Dev Biol 46:167–191, 1975.
27. Trisler GD, Schneider MD, Nirenberg M: Proc Natl Acad Sci USA 78:2145–2149, 1981.
28. Constantine-Paton M: In Subtelny S, Green PB (eds): "Developmental Order: Its Origin and Regulation." New York: Alan R. Liss, Inc., 1982, pp 317–349.
29. LeVay S, Hubel DH, Wiesel TN: J Comp Neurol 159:559–576, 1975.
30. Willshaw DJ, von der Malsburg C: Proc Roy Soc Lond B 194:431–445, 1976.
31. Whitelaw VA, Cowan JD: J Neurosci 1:1369–1387, 1981.
32. Straznicky C, Tay D, Hiscock J: Neurosci Lett 19:131–136, 1980.
33. Levine R, Jacobson M: Brain Res 98:172–176, 1975.
34. Law P: Thesis, Princeton University, Princeton, 1982.
35. Fawcett JW, Willshaw DJ: Nature 296:350–352, 1982.
36. Straznicky C, Tay D: J Comp Neurol 206:119–130, 1982.
37. Fraser SE, Ide C, Meyer RL: Science 221:293–295.
38. Rodieck RW: "The Vertebrate Retina." San Francisco: W.H. Freeman, 1973.
39. Arnett DW: Exp Brain Res 32:49–53, 1978.
40. Arnett D, Spraker TE: J Physiol Lond 317:29–47, 1981.
41. Mastronarde DN: J Neurophysiol 49:303–324, 1983.
42. Mastronarde DN: J Neurophysiol 49:325–349, 1983.
43. Mastronarde DN: J Neurophysiol 49:350–365, 1983.
44. Maturana HR, LeHuin JY, McCulloch WS, Pitts WH: J Gen Physiol Suppl 43:129–175, 1960.
45. Hebb DO: "The Organization of Behavior." New York: John Wiley & Sons, 1949 pp 60–78.

46. Changeux J-P, Danchin A: Nature 264:705–712, 1976.
47. Marr D: J Physiol (Lond) 202:437–470, 1969.
48. Reh T, Constantine-Paton M: J Neurosci :5, 1985.
49. Van Essen DC: In Spitzer NC (ed): "Neuronal Development." New York: Plenum Press, 1982, pp 333–371.
50. Srihari T, Vrbova G: J Neurocytol 7:529–540, 1978.
51. Thompson W, Kufer DR, Jansen JKS: Neuroscience 4:271–281, 1979.
52. Duxson MJ: J Neurocytol 11:395–408, 1982.
53. O'Brien RAD, Ostberg AJC, Vrbova G: J Physiol 282:571–582, 1978.
54. Brown MC, Holland RL, Hopkins WG: Ann Rev Neurosci 4:17–42, 1981.
55. O'Brien RAD, Ostberg AJ, Vrbova G: Neuroscience 5:1367–1379, 1980.
56. Stent GS: Proc Natl Acad Sci USA 70:997–1001, 1973.
57. Norden JJ, Constantine-Paton M: Soc Neurosci Abstr 8:450, 1982.
58. Murray M, Sharma S, Edwards MA: J Comp Neurol 209:374–385, 1982.
59. Nudell BM, Grinell AD: J Neurosci 3:161–176, 1983.
60. Brown ME, Jansen JKS, Van Essen D: J Physiol (Lond) 261:387–422, 1976.
61. Varon S, Adler R: Curr Top Devel Biol 16:207–252, 1980.
62. Turner JE, Barde Y, Schwab ME, Thoenens H: Devel Brain Res 6:77–83, 1983.
63. Pittman RN: Soc Neurosci Abstr 9:5, 1983.
64. Taylor AC, Kollros JJ: Anat Rec 94:7–28.

Neurobiology: Molecular Biological Approaches to
Understanding Neuronal Function and Development 169–177 (1985)

Structure, Function, and Expression of Calmodulin Genes

Anthony R. Means, John A. Putkey, Rosalia C.M. Simmen, and Martin W. Berchtold

Department of Cell Biology, Baylor College of Medicine, Houston, Texas 77030

Changes in the level of Ca^{2+} have been reported to be required in regulation of many key processes in the nervous system. In addition to overall cell metabolism, cyclic nucleotide levels, protein phosphorylation, and neurotubule as well as neurofilament function are Ca^{2+}-dependent [1–3]. Calcium also controls several activities that are specific to the nervous system. This ion depresses nerve membrane excitability [4], stimulates K^+ conductance [5], and is involved in excitation–secretion coupling, both at peripheral neuromuscular junctions [6] and central synapses [7]. Electrical synapses are uncoupled by Ca^{2+} as a result of increased permeability of gap junctions [8]. Calcium also modulates neurotranmsmitter synthesis [9] and is required for intracellular movement of substances by axonal transport [10].

LOCALIZATION OF CALMODULIN IN THE NERVOUS SYSTEM

It is now recognized that many Ca^{2+}-dependent processes in the nervous system are mediated through the ubiquitous intracellular Ca^{2+} receptor, calmodulin [2,3,11]. Calmodulin and a calmodulin-dependent protein kinase are major protein constituents of postsynaptic densities. We have utilized immunoelectron microscopy to localize calmodulin in rat cerebellum [12]. In Purkinje and granular cell bodies, calmodulin reaction product was found localized both on free ribosomes and on those attached to rough endoplasmic reticulum (RER) and the nuclear envelope. No calmodulin was observed in the cisternae of RER or the Golgi apparatus. Calmodulin did not appear to be concentrated in the soluble fraction of the cell under the conditions used. Rather, peroxidase reaction product could be seen associated with membranes of the Golgi apparatus, the smooth endoplasmic reticulum (SER), and the plasma membrane of both cell bodies and neuronal processes. In the neuronal dendrites, calmodulin appeared to be concentrated on membranes of the SER, small vesicles, and mitochon-

Received November 26, 1984.

dria. Also, granular calmodulin was observed in the amorphous material. In the synaptic junction, a large amount of calmodulin was seen attached to the inner surface of the postsynaptic membrane, whereas very little was observed in the presynaptic membrane or vesicles. These observations suggest that calmodulin is synthesized on ribosomes and discharged into the cytosol, and that it then becomes associated with a variety of intracellular membranes. Calmodulin also seems to be transported via neuronal processes to the postsynaptic membrane.

The sites of positive reaction product were confirmed by radioimmunoassay [12]. These data, together with many biochemical studies, reveal that calmodulin modulates a large number of events important in regulation of nervous system function [2,3,11]. It is likely that calmodulin mediates most (if not all) of the known functions of Ca^{2+} in mammalian brain by acting as an intracellular receptor for this regulatory ion. Thus, changes in the activity of calmodulin would be dictated by changes in the net flux or intracellular distribution of Ca^{2+} promoted by substances that regulate chemical and electrical synaptic transmission. One would predict that changes in the level of calmodulin would also have profound effects on these processes. Unfortunately, calmodulin is generally thought to be a constitutive component of most cells [11,13,14]. Therefore, elevation of calmodulin levels requires microinjection of the protein, transformation of the cells, or introduction of extra copies of the calmodulin gene under control of a regulatable promoter.

CALMODULIN AND CELL PROLIFERATION

To date, calmodulin levels have been shown to be altered by only three cellular events. These events are: 1) transformation of cells by oncogenic viruses, chemical carcinogens, or hormones; 2) at the G_1/S boundary of exponentially growing cells; and 3) during the response of a quiescent population of cells to a mitogenic stimulus. Calmodulin and calmodulin mRNA levels are 2–10 fold higher in several transformed cells relative to their non-transformed counterparts [15–17]. This increase is regulated at the level of synthesis and not degradation and occurs regardless of the nature of the transforming agent [16,17]. Thus, elevated levels of calmodulin, determined by radioimmunoassay, have been reported in cells transformed by RNA and DNA oncogenic viruses, chemical carcinogens, or hormones [11]. Calmodulin has also been demonstrated to be synthesized in a cell-cycle-dependent manner. Chafouleas et al [18] found that calmodulin is increased by a factor of two at the G_1/S boundary of CHO cells. Moreover, treatment of these cells in exponential growth with the calmodulin binding drug W13 results in the selective but reversible inhibition of the progression through S-phase. Further experiments revealed that this inhibition of DNA synthesis was due to a single W13-induced block at the G_1/S boundary. The drug had no effect on progression through G_1, G_2, or M.

Calmodulin levels are also regulated following a stimulation that induces quiescent cells to re-enter the cell cycle [19]. Calmodulin levels increase as cells exit the proliferative cycle and enter a quiescent or plateau stage. Following a mitogenic stimulus, CaM levels in Chinese hamster ovary (CHO) cells decrease by 50% within the first 60 min, remain at this level for 2–4 hr, and then increase prior to initiation of DNA synthesis. Changes in CaM mRNA precede changes in CaM protein, both with respect to the decrease as well as the subsequent increase. On the other hand, actin mRNA levels do not decrease nor does total poly A^+ RNA. These data suggest that at least the acute decrease is selective for CaM. The calmodulin binding drug W13 prevents cells from re-entering the cell cycle and results in two distinct and separable blocks, as determined by flow cytometry. The first block is at the plateau-G_1 transition, and the seconnd block occurs in early S-phase. Thus, cells that escape the first block accumulate at the second one. Both drug-induced blocks are reversible and cause no decrease in cell viability.

These data are consistent with previous studies, which have reported the existence of either two [20] or three [21] steps in the G_0–G_1–S transition. In addition, recent studies performed on serum-deprived 3T3 cells have demonstrated that reentry into the cell cycle can be broken down into at least two distinct steps [22–25]. The first is induction of competence. Cells that have been rendered competent by such mitogenic agents as platelet-derived growth factor are primed for the next step, progression into S-phase. Progression is mediated through other growth factors and can only occur in competent cells. Calmodulin may be involved in both of these steps.

CALMODULIN AND DNA REPAIR

Recent experimental results can be interpreted to suggest that the increase in calmodulin at G_1/S is important for optimal DNA repair prior to replicative DNA synthesis [26]. Bleomycin is a drug known to cause DNA damage by defined pathways. Concentrations of this drug can be readily determined that will result in potentially lethal damage to tissue culture cells. Under such conditions, approximately 90% of the cells are killed. The remaining 10%, however, can recover from the drug by repairing DNA and will eventually repopulate the culture dish. If cells are selected for potentially lethal damage and released from Bleomycin into the presence of media containing W13 (but not W12), all the cells are killed within 3 hr. The most obvious explanation of these results is that calmodulin is required for DNA repair. When calmodulin is neutralized with W13, then the cells cannot repair the DNA damage that resulted from Bleomycin. This results in cell death during the subsequent replicative phase.

A more direct approach to the postulate that calmodulin might be involved in DNA repair was to examine the effects of W13 on recovery from Bleomycin-induced damage by the nucleoid sedimentation assay described by Durkacz et al [27]. Nucleoids are defined as nuclei treated with high salt to remove the histones. Dispersion of the chromatin results in a characteristic sedimentation when centrifuged through a 15–30% sucrose gradient. Damage to DNA results in a more dispersed chromatin, and consequently the nucleoids sediment more slowly. The distance migrated between nucleoids from control and experimentally damaged cells is a measure of the degree of DNA breakage. The rate and extent of recovery from such damage is related to the ability of the cells to repair the DNA. W13 completely prevents recovery from Bleomycin-induced DNA damage, providing further evidence that calmodulin may be involved in the repair process [26].

ISOLATION AND CHARACTERIZATION OF THE CHICKEN CALMODULIN GENE

The experiments summarized above yielded data to suggest that cellular CaM levels could be altered either by transformation or during the cell cycle at the G_1/S boundary. Since the transformed cell types used in our study had a shorter cell cycle time than their nontransformed counterparts [16], it was possible that the primary intracellular regulation of CaM was cell-cycle-dependent, regardless of the genotype.

In order to continue our quest for mechanisms controlling the cell CaM level, we turned our attention to the cloning of CaM cDNAs and genes. Because of the high concentration of CaM in the electroplax of the electric eel, this tissue was utilized as a source of CaM mRNA. The first cloned cDNA encoded the C-terminal 35% of CaM (amino acids 93–148) and 182 nucleotides (NT) of 3′ noncoding region [28]. This recombinant molecule (pCM109) was next used to select a clone that was demonstrated by DNA sequencing to be full length and contain 26 NT of 5′ noncoding region [29]. The noncoding region contained two polyadenylation signals 280 NT apart that both gave rise to functional mRNAs. It was subsequently concluded that a single 5,500 NT primary transcript was utilized to produce three functional mRNA molecules 2,000, 1,100, and 820 NT in length.

The full length eel clone pCM116 was utilized to select CaM sequences from a cDNA library constructed in λ phage. The overlapping cDNA clones were sequenced and found to contain the entire coding region of chicken CaM, together with 94 NT of 5′ nontranslated and 857 NT of 3′ nontranslated regions [30]. The chicken cDNA clones were utilized to isolate CaM gene sequences from a bacteriophage λ library containing chicken DNA [30]. The initial isolate contained 13.0 kb of DNA but was determined not to contain the 5′ end of the CaM gene. We have now determined the structural organization of the chicken chromosomal gene for CaM by analysis of overlapping genomic clones obtained from Charon 4A and cosmid DNA libraries [31]. These clones together span 39 kb of chicken genomic DNA. The entire length of the CaM gene is 12 kb and, as such, is 8.3-fold longer than the mature mRNA of 1,404 NT. Seven intervening sequences of highly variable size separate the structural sequences into eight segments. The smallest of these introns (intron A; 19 bp) divides the 5′ untranslated region, while the two largest introns, of approximately 4 kb each (introns B and C), separate the first 30 NT of the CaM structural sequences from the initiation codon ATG and the rest of the gene, respectively. The exon sequences, in contrast, are of relatively uniform size, except for exon VIII, which is 885 bp in length.

The pattern of interruption of the CaM gene by introns is of particular interest in light of the calcium binding properties of CaM. Each molecule of CaM can bind four molecules of calcium. Therefore, it was possible that the introns would be positioned at the boundaries of the putative functional calcium binding subdomains, as had been observed for other genes where functional domains of proteins are encoded in individual exons [32–34]. The positions of the introns in the CaM gene with respect to the protein structure are summarized as follows: B between the ATG initiation codon and aa 1; C between aa 10–11; D between aa 58–59; E between aa 82–83; F between aa 100–101, and G between aa 139–140. Thus, the intervening sequences D, F, and G interrupt the calcium binding subdomains II, III, and IV [2], whereas intron E occurs in the linker that serves to divide the N- and C-terminal halves of the protein. The location of intron E is compatible with the theory that the primary structure of CaM has evolved by intragenic duplication of primordial sequences [35]. However, the functional significance of the location of the other introns in relation to the evolutionary pressure to conserve the sequences in the calcium binding subdomains is as yet unclear. The absence of any obvious correlation between the positions of the introns relative to the structural domains of the protein is not unique to the CaM gene; the exons of the rat preprocarboxypeptidase A gene [36] and the chicken fibronectin gene [37] also do not correspond in location to the known functional domains of the molecule.

DNA sequencing of the 5′ end of the CaM gene and reverse transcription of the cDNA from the 5′ noncoding region identified the presumptive CAP site of the mRNA to be nucleotide A, located 103 bp 5′ from the initiator codon ATG. This is very similar to the 96–98 NT 5′ nontranslated region of the Xenopus CaM gene, deduced from primer extension experiments by Chien and Dawid [38]. The unique band observed in our primer extension studies suggests that transcription of the CaM gene starts at a single site. The characteristic consensus sequences present in most eukaryotic genes—namely, the Goldberg-Hogness promoter (TATA box) and the "CAAT" sequence—were localized 29 and 58 bases, respectively, 5′ from the transcriptional initiation site of this gene. In addition, a conserved element that has been implicated in promoter function in simian virus 40 [39] was found at two different positions 5′ to the cap site in the CaM gene. In each case, these repeats (at −157 to −152 and −200 to −195 bp) agreed with the consensus sequence CCGCCC in all but one postion. The 5′ noncoding region is interrupted by a small intron of 19 bp. Although interruptions of the 5′ untranslated region of genes are fairly common [40–42], the very small size of the intron within this region appears to be less so.

More than 200 nucleotides flanking the 5′ end of the CaM gene have been sequenced. Computer-assisted analysis of the DNA sequences 5′ from the CAP site revealed no obvious homology between the 5′ flanking regions of the CaM gene and of the gene coding for another calcium binding protein, myosin light chain 3 [43]. However, the overall organization of these two chicken genes is strikingly similar. Both genes have multiple introns of varying size dividing the structural sequences into small exons. The intron/exon ratio for both genes is relatively high. In addition, the longest exon in both genes spand the 3′ untranslated region. Finally, the initiator codon ATG in both genes is separated from the first nucleotide of the structural sequences by a very large intron (approximately 4 kb). Further comparisons of the chicken CaM gene structure with the limited data available for the structure of the genes encoding an ectodermal specific calcium binding protein in sea urchin [44], a myosin light chain in rat [45], and troponin C in quail [46] indicate similarities in the location of introns within the calcium binding domains. These data taken together suggest a possible conservation of genomic structure related to the calcium binding and regulatory domains of calcium binding proteins.

ISOLATION OF A CHICKEN CALMODULIN-LIKE GENE

A second calmodulin-related DNA sequence was obtained by screening the chicken genomic library with the partial eel clone pCM109 under relaxed conditions (60°C; 6 × SSC). This molecule, CM-1, represents a single copy gene that encodes a 148-amino acid protein that bears an 87% sequence homology with chicken calmodulin [47]. This gene does not contain intervening sequences but does possess a TATA box, an ATG initiation codon, is in open reading frame, and has a single polyadenylation signal 486 NT from the TGA termination codon. This gene is flanked both 5′ and 3′ by a nine-base pair direct repeat. These properties are similar to those attributed to so-called processed genes. If this is true, CM-1 would represent a reverse transcribed copy of the CL-1 processed transcript. This copy, lacking introns, would be inserted back into the genome. Hybridization analysis of poly A^{+}RNA reveals the presence of an mRNA in several chicken tissues that is complementary to CM-1 [47]. Such data are compatible with the tissue-specific expression of a processed calmodulin gene and predict the presence of a calmodulin-like protein that differs in 19 of the 148 amino acids from calmodulin.

STRUCTURE/FUNCTION STUDIES OF CALMODULIN USING PROTEINS PRODUCED IN PROKARYOTIC EXPRESSION VECTORS

If the calmodulin-like gene CM-1 is expressed in a tissue-selective manner, the encoded protein would be the first calmodulin-like molecule to be demonstrated in a vertebrate species. As a prelude to answering this question, it was decided to develop recombinant DNA molecules capable of synthesizing the protein encoded by CM-1. Methods developed to purify the protein from bacterial extracts could then be applied directly to chicken tissues. In addition, an antibody could be produced that would selectively recognize CM-1 protein (but not authentic calmodulin) and used to localize this molecule in muscle as well as in other tissues and cells. Even if CM-1 protein did not exist in vivo, the bacterial produced protein would be useful to probe which domains of calmodulin are involved in which of its many proposed functions. The protein encoded by CM-1 would have amino acid substitutions in Ca^{2+} binding domains I, II, and IV [47]. By taking advantage of common restriction endonuclease sites in the three naturally occurring nucleic acids available for use [29,30,47], we have engineered three additional calmodulin-like proteins with specific changes introduced into a) domain I; b) domain II; and C) domains II and IV. Finally, bacterial expression vectors will allow us to metabolically label calmodulin or the calmodulin-like proteins with any isotope or radioactive amino acid desired. In this manner, direct binding studies of unmodified proteins can be carried out in vivo, in vitro, or using the gel overlay procedure.

A strategy was devised to construct an expression vector for calmodulin using the chicken cDNA [30], a pUC8 vector [48], and a Tac promoter [49] as the DNA components. The recombinant molecule was engineered so that the ATG was 7 NT from the ribosome biding site of the lac operon component of Tac [50]. A colony was selected after introduction into a bacterial host that produces 3% of its total protein as calmodulin. The bacterially produced calmodulin appears to be identical to rat testis calmodulin by the following criteria: 1) amino acid composition; 2) heat stability; 3) migration on one- and two-dimensional denaturing gels, both in the absence or presence of Ca^{2+}; 4) crossreactivity in a specific calmodulin radioimmunoassay; and 5) immunoprecipitation with a polyclonal affinity-purified calmodulin IgG.

The CM-1 gene was inserted into the calmodulin expression vector in such a fashion that the ATG and first ten amino acids are encoded by chicken calmodulin, whereas the remainder of the protein is encoded by CM-1 sequences [50]. A clone was selected in which again the level of expression was 3% of the total protein. The CM-1 protein was characterized with respect to rat testis and bacterially synthesized calmodulin. The CM-1 protein is similar to calmodulin in that it is heat stable, can be purified on phenothiazine-Sepharose, crossreacts in a calmodulin radioimmunoassay, and, in the presence of EGTA, migrates identically to calmodulin on 1D-SDS gels. CM-1 protein can be distinguished from calmodulin in that it undergoes an aberrant

Ca^{2+}-dependent mobility shift on SDS gels, is more basic than calmodulin upon analysis by 2D gel electrophoresis, and shows less enhancement of tyrosine fluorescence upon Ca^{2+} binding. Finally, CM-1 protein can be separated from calmodulin by chromatography on thiol-Sepahrose, since it contains 2-cys residues and calmodulin does not.

Ca^{2+} binding experiments reveal that CM-1 protein contains four functional sites but that the affinity of the two high affinity sites seems to be somewhat less than that for either rat testis CaM or bacterially synthesized CaM. The CM-1 protein will maximally activate bovine brain phosphodiesterase, but the K_m is three times greater than that for CaM. However, CM-1 protein will only partially activate the brain phosphatase calcineurin, although the K_m of activation is similar for that of CaM. These data suggest that the amino acid differences between the CM-1 protein and calmodulin have been selected to retain CaM-like (or at least Ca^{2+}binding protein-like) functions.

The question that naturally arises is what evolutionally pressure exists that would induce selective mutations in a gene that is apparently a CaM processed gene. Stein et al [47] showed that ^{32}P-labeled CM-1 DNA hybridized to an mRNA species present in chicken muscle. Therefore, the CM-1 gene could have undergone selective mutations and evolved to encode a protein that is expressed and performs a subset of CaM activities or has acquired new activities that are vital to the survival of the organism. If such a protein exists as a minor CaM species, it would have properties similar to bacterially synthesized CM-1 protein and be virtually indistinguishable from the bulk of the CaM by techniques that are usually employed to assess CaM. We are now exploiting the properties of the bacterially synthesized CM-1 protein in an effort to detect a similar protein in chicken tissue.

CHANGES IN CALMODULIN AND PARVALBUMIN DURING DEVELOPMENT

Parvalbumin (PV) belongs to a family of low molecular weight, high affinity Ca^{2+}-binding proteins that are related by their homologous primary sequences [51,52]. In contrast to the multifunctional Ca^{2+} receptor calmodulin (CaM), which is present in all eukaryotic cells [11,14], PV is only found in a limited number of vertebrate tissues, where it is restricted to a few distinct cell types [53]. In mammals, PV is most concentrated in fast-twitch muscle fibers and may be involved in the relaxation process [54,55]. Crossinnervation of fast skeletal muscle with a nerve normally connected to slow muscle leads to a considerable decrease of PV concentration, indicating that expression is dependent on neuromuscular interactions [56]. Parvalbumin has also been purified from rat brain but is present in much lower quantity than in fast skeletal muscle ($\approx 0.01\%$ vs 1.0% of the total protein, respectively) [57]. Immunohistochemical methods have demonstrated that distribution of PV in brain is restricted to a subpopulation of neurons [58]. However, neither the molecular basis nor the physiological significance of this cell-specific expression of PV is known.

We have isolated a rat PV cDNA clone and used this as a probe to examine changes in PV mRNA during muscle and brain devlopment [59]. A cDNA library was constructed in PUC8/PUC9 plasmid vectors from adult poly (A^+) RNA isolated from rat gastrocnemius muscle. The library was screened with a 17-mer oligonucleotide encoding amino acids 28–33 of rat PV. One recombinant (9f) was confirmed as a PV clone by DNA sequencing and was shown to contain 73% of the protein coding sequence. Hybridization of clone 9f to RNA separated by electrophoresis revealed two species 700 and 1,100 NT in length, but genomic blotting indicates that PV may be a single copy gene. Highest levels of PV mRNA are found in the gastrocnemius, which is a fast contracting/relaxing muscle. Skin contains the next highest amount of PV mRNA, followed in order by brain and the slow-twitch soleus muscle. Rat muscle PV mRNA levels increase 15–20 fold between postnatal days 4 and 20, as measured by dot blot hybridization of total RNA, whereas only a slight increase was observed when young and adult brains were compared. The changes in PV mRNA during development appear to be selective, since mRNA coding for the structurally homologous Ca^{2+} binding protein calmodulin was found to change only slightly in muscle. However, CaM mRNA levels decrease during the early days of brain ontogeny. Thus, the mRNAs that encode the homologous Ca^{2+}-binding proteins PV and CaM appear to be developmentally regulated in a tissue-specific manner.

ACKNOWLEDGMENTS

The research discussed in this review was supported in part by the following research grants to A.R.M.: NIH grant GM-33976, American Cancer Society grant BC-326, Robert A. Welch Foundation grant Q-611, and The Jerry Lewis Neuromuscular Disease Research Center of the Muscular Dystrophy Association. J.A.P. was supported by a Damon Runyon–Walter Winchell Cancer Fund Fellowship, no. DRG-582. M.W.B. was supported by a Swiss National Foundation fellowship for young advanced scientists, no. 83.011.082.

REFERENCES

1. Dedman JR, Brinkley AR, Means AR: Adv Cyclic Nucleotide Res 11:131–174, 1979.
2. Cohen P: Nature 296:613–620, 1982.
3. Nestler EJ, Greengard P: Nature 304:583–588, 1983.
4. Frankenhaeuser B, Hodgkin AL: J Physiol (Lond) 137:218–244, 1957.
5. Meech RW: J Physiol (Lond) 237:259–277, 1974.
6. Katz B, Miledi R: J Physiol (Lond) 203:459–487, 1969.
7. Douglas WW, Kanno T, Sampson S: J Physiol (Lond) 191:107–113, 1967.
8. Loewenstein WR: Cold Spring Harbor Symp Quant Biol 40:49–63, 1976.
9. Yamauchi T, Fujisawa H: Biochem Biophys Res Comm 90:28–34, 1979.
10. Brady ST, Lasek RJ: J Cell Biol 89:607–614, 1981.
11. Means AR, Dedman JR: Nature 285:73–77, 1980.
12. Lin CT, Dedman JR, Brinkley BR, Means AR: J Cell Biol 85:473–480, 1980.
13. Means AR: Rec Prog Horm Res 37:333–368, 1981.
14. Means AR, Tash JS, Chafouleas JG: Physiol Rev 62:1–38, 1982.
15. Watterson DM, van Eldik LJ, Smith RE, Vanaman TC: Proc Natl Acad Sci USA 73:2711–2715, 1976.
16. Chafouleas JG, Pardue RL, Brinkley BR, Means AR: Proc Natl Acad Sci USA 78:996–1000, 1981.
17. Zendegui JG, Zielinski RE, Watterson DM, van Eldik LJ: Mol Cell Biol 4:883–889, 1984.
18. Chafouleas JG, Bolton WE, Hidaka H, Boyd AE, Means AR: Cell 28:41–50, 1982.
19. Chafouleas JG, Lagace L, Bolton WE, Boyd AE, Means AR: Cell 36:73–81, 1984.
20. Riddle VGH, Pardee AB: J Cell Physiol 103:11–15, 1980.
21. Jonah GJ, Baserga R: J Cell Physiol 105:347–354, 1980.
22. Pledger WJ, Stiles CD, Antoniades HN, Scher CD: Proc Natl Acad Sci USA 74:4481–4485, 1977.
23. Stiles CD: Cell 33:653–655, 1983.
24. Stiles CD, Capone GT, Scher CD, Antoniades HN, VanWyk JJ, Pledger WJ: Proc Natl Acad Sci USA 76:1279–1283, 1979.
25. Cochran BR, Rettel AC, Stiles CD: Cell 33:939–947, 1983.
26. Chafouleas JG, Bolton WE, Means AR: Science 224:1346–1348, 1984.

27. Durkacz BW, Irwin J, Shall S: Eur J Biochem 121:65–69, 1981.
28. Munjaal RP, Chandra T, Woo SLC, Dedman JR, Means AR: Proc Natl Acad Sci USA 78:2330–2334, 1981.
29. Lagace L, Chandra T, Woo SLC, Means AR: J Biol Chem 258:1684–1688, 1983.
30. Putkey JA, Ts'ui KF, Tanaka T, Lagace L, Stein JP, Lai EC, Means AR: J Biol Chem 258:11864–11870, 1983.
31. Simmen RCM, Tanaka T, Ts'ui KF, Putkey JA, Scott MJ, Lai EC, Means AR: J Biol Chem 260:907–912.
32. Stein JP, Catterall JF, Kristo P, Means AR, O'Malley BW: Cell 21:681–687, 1980.
33. Eaton WA: Nature (Lond) 284:183–185, 1980.
34. Heinrich G, Kronenberg HM, Potts JT Jr, Habener JF: J Biol Chem 259:3320–3329, 1984.
35. Iida Y: J Mol Biol 159:167–177, 1982.
36. Quinto C, Quiroga M, Swain WF, Nikovits WC Jr, Standring DN, Pictet RL, Valenzuela P, Rutter WJ: Proc Natl Acad Sci USA 79:31–35, 1982.
37. Hirano H, Yamada Y, Sullivan M, De Crombugghe B, Pastan I, Yamada KM: Proc Natl Acad Sci USA 80:46–50, 1983.
38. Chien Y, Dawid IB: Mol Cell Biol 4:507–513, 1982.
39. Fromm M, Berg P: J Mol Appl Genetics 1:457–481, 1982.
40. Leicht M, Long GL, Chandra T, Kurachi K, Kidd VJ, Mace M Jr, Davie EW, Woo SLC: Nature (Lond) 297:655–659, 1982.
41. Yu-Lee L, Rosen JM: J Biol Chem 258:10794–10804, 1983.
42. Dugaiczyk A, Woo SLC, Colbert DA, Lai EC, Mace ML, O'Malley BW: Proc Natl Acad Sci USA 76:2253–2257, 1979.
43. Nabeshima Y, Fujii-Kuriyama Y, Muramatsu M, Ogata K: Nature (Lond) 308:333–338, 1984.
44. Carpenter CD, Bruskin AM, Hardin PE, Keast MJ, Anstrom J, Tyner AL, Brandhorst BP, Klein WH: Cell 36:663–671, 1984.
45. Periasamy M, Strehler EE, Garfinkel LI, Gubits RM, Ruiz-Opazo N, Nadal-Ginard B: J Biol Chem 259:13595–13604, 1984.
46. Maisonpierre PC, Emerson CP Jr: J Cell Biochem Supp 8B, 49 (abstract), 1984.
47. Stein JP, Munjaal RP, Lagace L, Lai EC, O'Malley BW, Means AR: Proc Natl Acad Sci USA 80:6485–6489, 1983.
48. Vieira J, Messing J; Gene 19:259–268, 1982.
49. Russell DR, Bennett GN: Gene 20:231–243, 1982.
50. Putkey JA, Slaughter GR, Means AR: J Biol Chem (in press).
51. Kretsinger RH: CRC Crit Rev Biochem 8:119–174, 1980.
52. Goodman M, Pechère JF, Haiech J, Demaille JG: J Mol Evol 13:331–352, 1979.
53. Berchtold MW, Celio MR, Heizmann CW: J Biol Chem 259:5189–5196, 1984.
54. Gillis JM, Thomason D, Lefèvre J, Kretsinger RH: J Muscle Res Cell Motil 3:377–398, 1982.
55. Heizmann CW, Berchtold MW, Rowlerson AM: Proc Natl Acad Sci USA 79:7243–7247, 1982.
56. Muntener M, Berchtold MW, Heizmann CW: Muscle and Nerve (in press).
57. Berchtold MW, Wilson KJ, Heizmann CW: Biochemistry 21:6552–6557, 1982.
58. Celio MR, Heizmann CW: Nature (Lond) 293:300–302, 1981.
59. Berchtold MW, Means AR: Proc Natl Acad Sci USA (in press).

Neurobiology: Molecular Biological Approaches to
Understanding Neuronal Function and Development 179–184 (1985)

Molecular Genetic Approaches to the Study of Expression of Opioid Peptide Genes

Edward Herbert, Michael Comb, Mitchel Martin, Dane Liston, Audrey Seasholtz, and James Douglass

Department of Chemistry, University of Oregon, Eugene, Oregon 97403

Neuropeptides have been shown to mediate diverse kinds of behavior in mammals, including responses to stress, pain, and anxiety. These peptides may function as neurotransmitters or neuromodulators in the nervous system or as hormones in the circulatory system. This explains why many neuropeptides are found in both the endocrine and nervous systems. After release from the neuron or endocrine gland, these peptides combine with receptors on secondary neurons or on target cells to initiate an intracellular cascade that activates the neuron or target cell. For a complete understanding of behavior, all of these events must be defined at the molecular level. This seemed like an impossible task just a few years ago. However, molecular biology has in the recent past provided us with the methods and the elemental knowledge to begin the analysis of these events.

Application of molecular biology to the study of the brain has led to some startling discoveries in the past five years. The sequencing of neuropeptide genes by recombinant DNA techniques has revealed the existence of a new type of protein in the brain, called the polyprotein, which serves as a source of several types of neuropeptides. From the structures of several polyproteins, it has been possible to predict the existence of many new types of neuropeptides [1]. This discovery has made us realize that the repertoire of neuropeptides in the brain is much larger than we had previously thought. Each polyprotein is expressed in several sites in the brain, where it may be processed to different peptides. The peptides derived from a single polyprotein may act on different target tissues to coordinate complex behavioral responses [1]. Thus, diversity in the production of neuropeptides can be generated at more than one level of gene expression.

The impact of molecular biology on brain research is no more apparent than in the study of opioid peptides. More than 16 different opioid peptides were discovered between 1975 and 1984. Met- and leu-enkephalin, the simplest of these peptides, were discovered first [2]. The other opioid peptides are C-terminal extensions of either met-enkephalin or leu-enkephalin [3]. The biosynthetic origin of these peptides was unclear until recombinant DNA research in several labs revealed the existence of three different opioid precursors: POMC (endorphin) [4–6], proenkephalin (enkephalin) [7–9], and prodynorphin [10]. All of these precursors contain multiple bioac-

Received November 26, 1984.

tive peptides. They are all expressed in a variety of sites in the nervous and endocrine systems, where they may be processed to different end products. POMC, for example, is processed to $ACTH_{1-39}$, β-lipotropin, and an N-terminal fragment in the anterior pituitary of the rodent. In the rodent neurointermediate lobe (NIL), each fragment is processed further to smaller peptides that have different actions than those released in the anterior pituitary [1].

Regulation of synthesis and release of peptides derived from POMC is also different in anterior and neurointermediate lobes. Release of anterior lobe peptides is stimulated by corticotropin releasing factor (CFR) and vasopressin [11,12]. Stimulation of release of these peptides by CRF and vasopressin can be prevented by glucocorticoids [13]. Release of peptides in the NIL is not affected by CRF or glucocorticoids [14], but instead is negatively controlled by dopaminergic compounds.

In primary cultures of rodent anterior pituitary, CRF stimulation of release of POMC peptides occurs within minutes of treatment. After prolonged periods of stimulation with CRF, the levels of POMC and mRNA start to increase and reach three to five times the control level after several days of continuous treatment with this regulator. An important question that arises is whether the two effects of CRF are coupled to one another by a common intracellular cascade and where glucocorticoids act in this cascade of events to inhibit the actions of CRF.

The mouse pituitary tumor cell line ($AtT\text{-}20_{D16v}$) is also an excellent model system for studying regulation of expression of the POMC gene. This cell line synthesizes large quantities of POMC (2–4% of the total protein) and processes this precursor in a fashion almost identical to that of mouse anterior pituitary cells [15]. Regulation of the expression of POMC peptides by CRF and glucocorticoids is also similar to that of anterior pituitary cells, except that the effect of CRF on release of POMC peptides is not as great as in normal cells.

The release and synthesis of proenkephalin may also be regulated differently in different tissues. In the bovine adrenal medulla, neural input regulates the release of enkephalin-containing peptides [16–18]. In cultures of chromaffin cells from bovine adrenal medulla, the levels of enkephalin peptides and proenkephalin mRNA are dramatically increased by treatment with agents that elevate cyclic AMP [19]. Neurotransmitters may control the release of enkephalin peptides via the same pathways. At present, we know very little about the mechanisms involved in the action of these agents. Do these agents trigger an intracellular cascade that activates the proenkephalin gene? If so, what is the mechanism of gene activation?

Advances in molecular biology in the past few years have provided us with the tools for studying genetic regulation of neuropeptides and their receptors. A number of neuropeptide genes, including those that code for the three classes of opioid peptides, have been cloned and sequenced. The availability of cloned genes of known sequence makes it possible to study genetic control of these genes by recently developed techniques of gene transfer. If the regions flanking the gene (5′ and 3′) are present in the cloned gene, one can identify nucleotide sequences that regulate transcription, such as promoters and enhancers (including steroid receptor binding sites) [20–22].

GENE TRANSFER EXPERIMENTS

The introduction of the opioid genes into an independent cell system in gene transfer experiments makes it possible to examine the role of specific DNA sequences in the expression of these genes. Since the mechanisms of gene expression differ markedly between eucaryotes and prokaryotes, the recipient cell system must be eucaryotic. We have selected two different in vivo eucaryotic systems as recipients for gene transfer experiments on opioid gene regulation. The Xenopus oocyte system has been used previously to transcribe and translate many different eucaryotic genes. The oocyte is an undifferentiated cell and will express, regulate, and process the introduced gene in a manner determined by its recognition of the control signals in the DNA sequence of the gene. This system has been used to locate three distinct transcriptional control signals within the 105 nucleotide residues preceding the mRNA cap site for the herpes thymidine kinase gene [23]. Studies with the sea urchin histone H_2A gene have also defined a region upstream from the TATA box that is involved in regulation of histone mRNA synthesis [24]. This system has several advantages over other systems for these types of mapping studies because 10–30% of the RNA produced in 24–48 hr by the oocyte will be transcripts complementary to the injected DNA [25]. Thus, the oocyte is a convenient and efficient assay system for studying the transcriptional activity of wild-type genomic sequences and DNA sequences that have been altered by site-specific mutagenesis in regions thought to be involved in regulation. By quantitation of the transcripts produced from the mutated genes, we can determine the exact DNA sequences involved in some of the transcriptional regulatory processes.

We have also performed gene transfer experiments in cultured cell lines to examine regulatory phenomena that may not be demonstrated in oocytes because of their undifferentiated nature. Cell lines have been selected that display glucocorticoid or polypeptide hormone regulation of the opioid genes. By using these cell lines, we should be able to localize additional regions of the opioid genes involved in regulation, especially regions required for steroid or polypeptide hormone regulation. These types of experiments have localized the regions of DNA necessary for glucocorticoid regulation in the mouse mammary tumor virus and human metallothionein genes. These methods have also been used to define transcriptional regulatory sites in the 5′ flanking regions of several other genes, including rabbit β-globin [26], insulin [27], and chymotrypsin [27]. They have also made it possible to study regulation of processing of precursor molecules to bioactive peptides as illustrated in the next section. By using these two complementary expression systems for gene transfer experiments, we hope to gain a better understanding of the role of specific DNA regulatory sequences (ie, promoters, enhancers, methylated CpG sites, palindromic regions, highly conserved sequences, etc) in the regulation of opioid gene expression.

THE EXPRESSION OF HUMAN PROENKEPHALIN IN MOUSE PITUITARY TUMOR CELLS THAT SYNTHESIZE AND PROCESS POMC

We have introduced and expressed the human proenkephalin gene in AtT-20 cells in order to study transcriptional regulation of this gene and to elucidate the mechanisms underlying processing of proenkephalin. The AtT-20 cell line has been used extensively as a model system to study the synthesis and processing of POMC, as pointed out before [28]. These cells possess well-developed secretory vesicles and store and release large amounts of ACTH and β-LPH. They also respond to physiological regulators in a manner similar to anterior pituitary cells [29]. In addition, AtT-20 cells chemically modify and proteolytically process, in a well-characterized step-wise fashion, POMC to ACTH and β-LPH [28]. Hence, these cells provide a well-characterized secretory pathway in which to examine the proteolytic processing and chemical modifications of neuropeptide precursors introduced by gene transfer.

In the brief description of results below we show that a recombinant plasmid carrying the human proenkephalin gene can be introduced and expressed in AtT-20 cells. The transfection of AtT-20 cells was performed by a co-transformation procedure [30] (see Fig. 1), using two kinds of plasmids, one containing the entire human proenkephalin gene with its promotor region and regulatory elements intact and the other containing the pRSVneo gene, which codes for a neomycin resistance factor [31]. Cells that take up the pRSVneo plasmid grow into colonies that are resistant to neomycin (see Fig. 1). Cells that do not take up and express this plasmid can be killed by antibiotic G419 (a neomycin derivative), providing a selection scheme for isolating transformed cells.

The results indicate that AtT-20 cells can be transformed to a G418-resistant phenotype using plasmid pRSVneo at stable transformation frequencies approximately ten-fold lower than previously reported for NIH/3T3 or CHO cells [31]. Southern analysis indicates that 50% of the G418-resistant colonies carried at least one copy of the human proenkephalin gene. Although the majority of the colonies contained between one and four copies of the gene, several colonies incorporated between ten and 30 gene copies. Northern blot analysis shows that several clones produce a 1.45 Kb RNA that hybridizes to proenkephalin cDNA. This is the same size proenkephalin mRNA produced by human pheochromocytoma tumors (adrenal medulla tumors). There appears to be no correlation between the number of integrated gene copies (gene dosage) and the level of proenkephalin mRNA transcripts found in the cell, as the clone A8-2 exhibits the highest levels of mature proenkephalin mRNA and contains the lowest gene copy number.

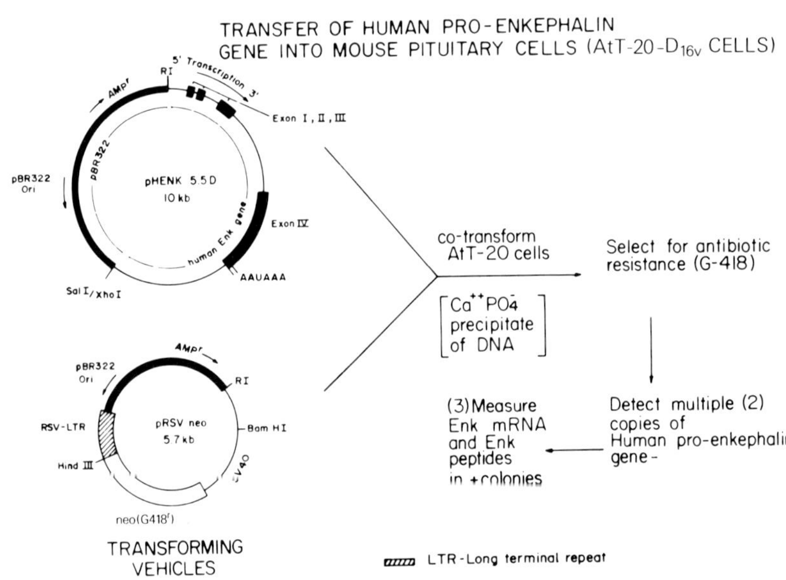

Fig. 1. Transfer of the human proenkephalin gene into mouse pituitary tumor cells (AtT-20-D_{16v} cells). The designation pHENK 5.5D stands for the whole plasmid containing the 5.5 Kb human proenkephalin gene.

The conclusion that the 1.45 Kb RNA transcript found in transformed AtT-20 cells is human and not mouse proenkephalin RNA is based on two observations. First, only the transformed AtT-20 clones exhibit the 1.45 Kb transcript, and second, the hybridization and wash conditions are stringent enough to greatly reduce cross hybridization to mouse proenkephalin mRNA, as determined by Southern analysis (most of mouse proenkephalin signal is removed by the final stringent wash).

Comparison of the level of proenkephalin mRNA produced by the AtT-20 clones to the levels of proenkephalin mRNA found in human pheochromocytoma and the levels of POMC found in the same RNA samples indicate proenkephalin transcripts range between 10 and 40% of the level of POMC transcripts, or about 0.3 to 1% of the total poly (A) RNA. This is a much higher level of expression than observed for pro-insulin cDNA, which has also been introduced into AtT-20 cells [32]. The high-level expression and correct size of human proenkephalin mRNA produced by AtT-20 cells indicate that the human proenkephalin promotor functions accurately and efficiently in the mouse AtT-20 cellular environment. Furthermore, since the plasmid introduced and expressed in these cells contained only sequences of pBR322 and human proenkephalin gene fragments, all of the proenkephalin control signals, such as the promotor, poly (A) addition sites, and RNA splicing signals, must be utilized by the recipient cells to express mature mRNA. Because pHENK 5.5 contains a 2.5 Kb deletion in intron III, it would appear that this DNA is not essential for correct RNA processing.

The results show that stable transformants of AtT-20 cells expressing both human proenkephalin mRNA and mouse POMC mRNA have been identified. Proenkephalin mRNA levels range between 20–50% of the levels of POMC mRNA, indicating high-level expression of the human proenkephalin gene in AtT-20 cells. We have also detected high levels of enkephalin products in transformed cells using leu- and met-enkephalin antibodies for radioimmunoassay of transformed cell extracts. The major immunoreactive product in these extracts has been identified as met- enkephalin by high performance liquid chromatography. Hence, AtT-20 cells are capable of performing all of the proteolytic cleavages of proenkephalin necessary to generate met-enkephalin.

Finally, it should be noted that expression of proenkephalin in bovine adrenal medulla chromaffin cells is regulated by neurotransmitters that act through cyclic-AMP-dependent mechanisms. Elevation of cAMP levels in chromaffin cells in culture by treatment of these cells with forskolin increases the level of proenkephalin mRNA several-fold. We have found that transformed AtT-20 cells that express human proenkephalin also respond to addition of cAMP by increasing the level of proenkephalin mRNA two- to three-fold. Thus, the region or regions of the proenekphalin gene that respond to regulation by cyclic AMP are present in transformed cells. It should be possible in the near future to identify these regulatory regions by deletion analysis and in vitro mutagenesis of the proenkephalin gene prior to gene transfer.

REFERENCES

1. Douglass J, Civelli O, Herbert E: Ann Rev Biochem 53:665–715, 1984.
2. Hughes J, Smith TW, Kosterlitz HW, Fothergill LA, Morgan BA, Morris HR: Nature 258:577–579, 1975.
3. Herbert E, Civelli O, Douglass J, Martens G, Rosen H: In: Litwack G (ed): "Biochemical Actions of Hormones," Vol. 12 (in press).

4. Roberts JL, Herbert E: Proc Natl Acad Sci USA 74:4826–4830, 1977.
5. Mains RE, Eipper BA, Ling N: Proc Natl Acad Sci USA 74:3014–3018, 1977.
6. Nakanishi S, Inove A, Kita T, Nakamura M, Chang ACY, Cohen SN, Numa S: Nature 278:423–427, 1979.
7. Noda M, Furutani Y, Takahashi H, Toyosato M, Hirose T, Inayama S, Nakanishi S, Numa S: Nature 195:202–206, 1982.
8. Gubler U, Seeburg P, Hoffman BJ, Gage LP, Udenfriend S: Nature 195:206–208, 1982.
9. Comb M, Seeburg PH, Adelman J, Eiden L, Herbert E: Nature 295:663–666, 1982.
10. Kakidani H, Furutani Y, Takahashi H, Noda M, Morimoto Y, Hirose T, Asai M, Inayama S, Nakanishi S, Numa S: Nature 298:245–249, 1982.
11. Vale W, Spiess J, Rivier C, Rivier J: Science 213:1394–1397, 1981.
12. Rivier C, Brownstein M, Spiess J, Rivier J, Vale W: Endocrinology 110:272–278, 1982.
13. Rivier C, Vale W: Endocrinology 113:1422–1426, 1983.
14. Rose PA, Policastro P, Herbert E: J Exp Biol 89:215–237, 1980.
15. Eipper BA, Mains RE: Endocrine Rev 1:1–27, 1980.
16. Schultzberg M, Lundberg JM, Hokfelt T, Terenius L, Brandt J, Elde RP, Goldstein M: Neurosci 3:1169–1186, 1978.
17. Lewis RV, Stern AS, Kilpatrick DL, Gerber LD, Rossier J, Stein S, Udenfriend S: J Neurosci 1:80–82, 1981.
18. Bohn MC, Kessler JA, Golightly L, Black IB: Cell Tissue Res 231:469–479, 1983.
19. Quach TT et al (1984) Mol Pharmacol (in press).
20. Buetti E, Diggelmann H: EMBO Journal 2:1423–1429, 1983.
21. Karin M, Haslinger A, Holtgreve H, Richards RI, Krauter P, Westpahl HM, Beato M: Nature 308:513–519, 1984.
22. Hynes N, Van Doyen AJJ, Kennedy N, Herrlich R, Ponta H, Groner B: Proc Natl Acad Sci USA 80:3637–3641, 1983.
23. McKnight SL, Kingsbury R: Science 217:316–324, 1982.
24. Grosschedl R, Birnstiel ML: Proc Natl Acad Sci USA 77:1432–1436, 1980.
25. Wickens MP, Woo S, O'Malley BW, Gurdon JB: Nature 285:628–634, 1980.
26. Dierks P, Van Ooyen A, Cochran MD, Dobkin C, Reiser J, Weissman C: Cell 32:695–706, 1983.
27. Walker MD, Edlund T, Boulet AM, Rutter WJ: Nature 306:557–561, 1983.
28. Herbert E: Trends Biochem Sci 6:184–188, 1981.
29. Herbert E, Allen RG, Paquette TL: Endocrinology 102:218–225, 1978.
30. Graham FL, VanDerEb AJ: Virology 52:456–467, 1973.
31. Gorman C, Padmanabhan R, Howard BH: Science 221:551–553, 1983.
32. Moore H-PH, Walker MD, Lee F, Kelly RB: Cell 35:531–538, 1983.

Index